T0296157

CAMBRIDGE PHYSICAL SERIES

VOLUMETRIC ANALYSIS

VOLUMETRIC ANALYSIS

BY

A. J. BERRY, M.A.
Fellow of Downing College, Cambridge

Cambridge :
at the University Press
1915

CAMBRIDGE
UNIVERSITY PRESS

University Printing House, Cambridge CB2 8BS, United Kingdom

Published in the United States of America by Cambridge University Press, New York

Cambridge University Press is part of the University of Cambridge.

It furthers the University's mission by disseminating knowledge in the pursuit of education, learning and research at the highest international levels of excellence.

www.cambridge.org
Information on this title: www.cambridge.org/9781107657137

© Cambridge University Press 1915

First published 1915
First paperback edition 2014

A catalogue record for this publication is available from the British Library

ISBN 978-1-107-65713-7 Paperback

PREFACE

THE present work has originated in connexion with teaching Volumetric Analysis to the author's students in the College Laboratory. Hitherto, most of the books on the subject available for the student have been elementary works, which do not attempt to cover much ground, and standard works such as that of Sutton. It was felt that the educative value of Volumetric Analysis justified the production of a book in which an attempt was made to treat the subject with some degree of thoroughness from the theoretical as well as from the practical point of view. In this connexion, particular care has been taken to make the conception of equivalent weights and normal solutions as clear as possible, and also to encourage the student to calculate his results from the actual chemical changes which take place, rather than to have recourse to the shorter method—excellent for the practical analyst—of always working with normal or deci-normal solutions, and calculating his results with the aid of "factors." The author ventures to hope that the inclusion of a chapter on the Theory of Indicators will increase the interest of the student in the theoretical part of the subject.

The author desires to express his most hearty thanks to Mr T. G. Bedford, Editor of the Cambridge Physical Series, for many valuable suggestions and criticisms both of the manuscript and of the proofs. His thanks are also due to Mr L. F. Newman for information regarding certain methods. In the preparation of the book, information on particular points has been freely sought from larger works, and the author would here express his gratitude to the Council of the British Association for their courtesy in permitting the reproduction of Figs. 5, 6 and 7, which are taken from Mr H. T. Tizard's valuable paper on the Sensitiveness of Indicators in the Report for 1911.

A. J. B.

DOWNING COLLEGE, CAMBRIDGE.
February, 1915.

CONTENTS

CONTENTS

CHAPTER I

INTRODUCTION

General principles

Methods employed in quantitative chemical analysis may be divided broadly into two main classes, gravimetric and volumetric. In the former, the constituents of a substance are determined by separation and weighed in the form of compounds of known composition. In the latter, the substance to be estimated is allowed to react in solution with another substance of which a solution of known strength has been made; and the volume of the solution of known strength which is required for the completion of the reaction with a certain definite volume of the solution of the substance to be determined is observed. In order that a volumetric determination may be successfully carried out, it is essential that the end of the reaction may be clearly visible to the eye by the appearance or disappearance of some characteristic colour in the solution. Volumetric analysis possesses a great advantage over gravimetric analysis, viz., that the determination of a substance may be carried out with a very much smaller expenditure of time. In a gravimetric method it is necessary to separate the particular constituent which it is desired to determine in a state of great purity; while in a volumetric method such perfect separation is very seldom required, the presence of relatively large quantities of other substances which do not interfere with the particular reaction having in general no effect upon the accuracy of the determination.

Accuracy of volumetric analysis

The conditions which determine the accuracy of a volumetric method are threefold: firstly the purity of the substance which is employed for making up the solution of definite strength known as the standard solution; secondly the accuracy of the measuring

B.

vessels; and thirdly, the sensitiveness of the change of colour or other device for indicating the completion of the reaction. When these three conditions are fulfilled, volumetric methods will be found to bear favourable comparison with the best gravimetric methods.

Equivalent weights

The equivalent weight of a substance is that weight of it which will react with a certain definite weight of some other substance. For example, it is an experimental fact that 169·9 grammes of silver nitrate will completely precipitate the chlorine in 74·56 grammes of potassium chloride by double decomposition according to the equation

$$AgNO_3 + KCl = AgCl + KNO_3.$$

Again, if silver nitrate be precipitated by means of hydrochloric acid, it is found that 169·9 grammes of silver nitrate will precipitate the chlorine in 36·47 grammes of hydrogen chloride. These experimental results are expressed by saying that one gramme molecular weight of silver nitrate is equivalent to 74·56 grammes of potassium chloride and to 36·47 grammes of hydrogen chloride respectively.

Again, it has been shown by experiment that one gramme molecular weight (40·01 grammes) of sodium hydroxide is capable of neutralizing exactly 49·04 grammes of sulphuric acid, or 36·47 grammes of hydrochloric acid, or 45·01 grammes of anhydrous oxalic acid. In other words, these particular weights of sulphuric, hydrochloric and oxalic acids are said to be chemically equivalent to one gramme molecule of caustic soda and to one another.

If we consider a third type of reaction, viz., the oxidation of a solution of ferrous sulphate in presence of dilute sulphuric acid to ferric sulphate by means of potassium dichromate, accurate experimental work has shown that one gramme atomic weight of iron (metal), or 55·85 grammes, requires 49·03 grammes of potassium dichromate for complete transformation from the ferrous to the ferric condition, the reaction taking place in accordance with the equation

$$6FeSO_4 + K_2Cr_2O_7 + 7H_2SO_4$$
$$= K_2SO_4 + Cr_2(SO_4)_3 + 7H_2O + 3Fe_2(SO_4)_3.$$

The equivalent weight of potassium dichromate is therefore 49·03 relative to one gramme atomic weight of iron undergoing oxidation from the ferrous to the ferric condition.

It is clear from what has been stated that the equivalent weight of a compound is not necessarily identical with its molecular weight although it is closely related to it. In the case of an element the equivalent weight is equal to the quotient of the atomic weight by the valency. Considering only substances which combine directly with hydrogen, we might define the equivalent of a substance as that weight of it in grammes which will combine with 1·008 grammes of hydrogen. But if we wish to extend our definition in a consistent manner to substances which do not combine directly with hydrogen, we must study the behaviour of such substances towards some other element whose equivalent weight relative to hydrogen is accurately known. For example, it has been shown that 1·008 grammes of hydrogen combine directly with 35·46 grammes of chlorine producing 36·47 grammes of hydrogen chloride. If this gas be now dissolved in water it will be found to neutralize 40·01 grammes of sodium hydroxide according to the equation

$$HCl + NaOH = NaCl + H_2O.$$

Again, the neutralization of sulphuric acid by caustic soda resulting in the formation of sodium sulphate and water takes place according to the equation

$$H_2SO_4 + 2NaOH = Na_2SO_4 + 2H_2O.$$

From this equation it follows that the equivalent weight of sulphuric acid is half its molecular weight. The double decomposition of hydrochloric acid and silver nitrate resulting in the formation of silver chloride and nitric acid shows that the equivalent of silver nitrate is 169·9.

In the third example which we have been considering, viz., the oxidation of a ferrous salt to the ferric condition by means of potassium dichromate, the value of the equivalent of this oxidizing agent is most easily seen from the fact that its decomposition may be regarded as due to the breaking down of the molecule $K_2Cr_2O_7$ into K_2O, Cr_2O_3, and three atoms of oxygen, which are effective in the conversion of the iron from the ferrous to the ferric condition.

Now it is an experimental fact that two molecules of hydrogen (four atoms) combine with one molecule (two atoms) of oxygen in the formation of water. Since the atomic weights of hydrogen and oxygen are respectively 1·008 and 16, it is clear that the equivalent of oxygen is 8, or in other words, one atom of oxygen is chemically equivalent to two atoms of hydrogen. Since potassium dichromate contains three atoms of oxygen which are available for the transformation of ferrous into ferric iron, these three atoms of oxygen are equivalent to six atoms of hydrogen. The equivalent weight of potassium dichromate is therefore one-sixth of its molecular weight or $\frac{1}{6}$ of 294·2 grammes or 49·03 grammes.

The three examples which we have discussed will indicate that by the adoption of equivalent weights a perfectly consistent inter-relationship between a large number of substances will be found to exist. We may therefore frame a definition of the term equivalent which will lead to such a relationship in the following terms. The equivalent of any substance, element or compound, is that weight of it in grammes which will either directly or indirectly bring one gramme of hydrogen into chemical action.

The importance of a consistent inter-relationship between the equivalent weights will be more readily apparent later. For the present the rôle of the equivalent in the calculation of results of volumetric determinations will be discussed.

Calculation of results

Suppose that a volume of v_1 cubic centimetres of a substance A is being determined, and that a volume of v_2 cubic centimetres of a substance B containing w grammes per cubic centimetre is required to complete the reaction, the weight x, in grammes, of substance A in each cubic centimetre is determined by the equation

$$\frac{v_1 x}{v_2 w} = \frac{\text{equivalent of } A}{\text{equivalent of } B} \quad \cdots\cdots\cdots\cdots (1).$$

Normal solutions

In volumetric analysis a considerable saving of arithmetical work may be effected by the employment of standard solutions which contain the equivalent weight of the substance in grammes dissolved in one litre of the solution. Such standard solutions are termed normal solutions; or in other words, a normal solution may be defined as a solution of such a strength that one litre of it contains that weight of the solute which is chemically equivalent to one gramme of available hydrogen. For many purposes solutions of normal strength are too strong; in such cases, it is usual to employ solutions of semi-normal or of deci-normal strength, while for certain special work solutions of centi-normal strength are used. Solutions of normal, semi-normal, deci-normal, and centi-normal strength are conveniently abbreviated by the symbols $N, \frac{N}{2}, \frac{N}{10}, \frac{N}{100}$ respectively. The advantage of employing solutions of normal or of a sub-multiple of normal strength may be seen from the fact that 20 c.c. of a solution of deci-normal HCl will neutralize 20 c.c. of a deci-normal solution of NaOH or will precipitate 20 c.c. of a deci-normal solution of $AgNO_3$, in every case without any residue of either reagent remaining unacted on.

The reaction between equal volumes of solutions of normal or of some sub-multiple of normal strength however does not hold good in all cases, as the following example will show. Potassium bi-iodate $KH(IO_3)_2$ is a substance which can react either as an acid or as an oxidizing agent. In the former case, one molecule of the substance will neutralize one molecular proportion of potassium hydroxide with formation of two molecules of potassium iodate according to the equation

$$KH(IO_3)_2 + KOH = 2KIO_3 + H_2O.$$

This reaction indicates that a normal solution of potassium bi-iodate should contain the molecular weight of the substance (390 grammes) dissolved in one litre. In the latter case potassium bi-iodate will liberate iodine from potassium iodide in presence of an acid, one molecule of the bi-iodate liberating six molecules of iodine as represented by the equation

$$KH(IO_3)_2 + 10KI + 11HCl = 6I_2 + 11KCl + 6H_2O.$$

Since one molecule of potassium bi-iodate liberates six molecules of iodine which are equivalent to twelve atoms of hydrogen, it follows that as an oxidizing agent the normal solution of this substance should contain one-twelfth of the molecular weight in grammes per litre.

It is clear, therefore, that the equivalent weight of a volumetric reagent is not an invariable magnitude like the molecular weight, but may be different in different reactions. It is in other words impossible to prepare normal solutions of *all* substances which shall possess the property that a given volume of one shall react quantitatively with an equal volume of any of the others. However, the number of substances which do conform consistently to the normal system is so great that even at the present time the system is of great practical value.

For many technical purposes it is usual to prepare standard solutions of such strength that the number of cubic centimetres of solution required for titration corresponds to a certain percentage of purity of the substance which is being examined.

Classification of methods in volumetric analysis

There are three main methods in volumetric analysis. The first is the direct method which includes all cases where the substance to be determined is estimated as the result of a single decomposition in the solution; such processes include the determination of acids by means of alkalis, of chlorides by silver nitrate, and of ferrous salts by means of potassium dichromate as well as numerous others. Secondly, there are indirect methods in which one or more intermediate reactions come into play; such processes include the determination of peroxides by distilling with hydrochloric acid, passing the chlorine into excess of potassium iodide, and determining the liberated iodine by means of sodium thiosulphate. Lastly, there are methods in which the substance to be determined is treated with a measured excess of some other substance for the purpose of reacting with it, and the excess of the added substance is then determined by some other reagent. As an example of this residual method, reference may be made to the estimation of ammonium salts by adding a known excess of standard alkali, boiling the solution to effect the decomposition of the ammonium salt and the removal of the ammonia, and then determining the excess of alkali remaining over by means of standard acid.

There are no general rules for selecting any one of these three general methods in preference to the others. There are many

substances which may be determined by any of these methods
with equally satisfactory results. In certain cases it is convenient
to make a combination of the direct or of the indirect method
with the residual method.

Besides classifying volumetric methods according to the
general experimental procedure, it is convenient to classify them
according to the type of reaction which takes place in effecting
the estimation. Three main types of chemical action are made
use of in most of the commonly occurring volumetric processes.
Firstly there is neutralization, or the double decomposition of acids
and bases resulting in the formation of a salt and water; this method
is employed for the estimation of acids and alkalis. In the second,
analysis is effected by oxidation or reduction, the substance being
converted from a lower to a higher degree of oxidation by means
of an oxidizing agent of known value, or conversely from a higher
to a lower degree of oxidation by means of a standard reducing
agent. In the third the determination of the substance is
effected by precipitation in an insoluble form by double decomposition, as in the determination of silver in solution by potassium
chloride.

Classification of volumetric processes according to the main
types of reaction which take place is convenient for many purposes,
but it does not include all processes. The determination of copper
by the decolorization of an ammoniacal solution of the metal
by means of potassium cyanide does not admit of classification
under any of the foregoing heads.

Apparatus employed in volumetric work

It is perhaps superfluous to give an account of the ordinary
measuring vessels—pipettes, burettes, and measuring flasks—
employed in volumetric analysis, since no written description of
them can possibly make the reader as familiar with them as
practical work in the laboratory. A few details regarding the use
of burettes and pipettes however must be given. These vessels
must be thoroughly cleansed before use, and then washed out
with a small quantity of the liquid with which they are to be filled
in order to prevent dilution of the solutions with water adhering

to the surface of the glass. The correct reading of graduated apparatus is important. In consequence of capillarity, the surface of a liquid in a narrow vessel is always curved; and if, as is usually the case, the liquid wets the glass, the surface will take the form of a concave meniscus. In reading the height of the liquid, the level of the lowest part of the meniscus is always taken, and it is important to place the eye on a level with the meniscus, as otherwise errors will be introduced. In allowing the contents of pipettes to flow into flasks previous to titration, sufficient time must be allowed for the liquid to drain down the walls of the vessel. The same remark may be made in titrating liquids when the liquid is allowed to flow rapidly from the burette; before reading the level of the liquid, time must be given for the liquid to drain down the tube, since otherwise the burette reading will be too high.

Errors in volumetric analysis

It has been already stated that the accuracy of a volumetric process depends on three main factors, viz., the purity of the substance employed for making up the standard solutions, the accuracy of the measuring vessels, and the sensitiveness of the change of colour for indicating the completion of the reaction. Besides these factors there are others which depend upon the conditions of experiment, and in this connexion reference must be made to the influence of the strength of the working solutions upon the accuracy of the process. This may perhaps best be illustrated by a practical example. Suppose it is desired to determine the percentage of sodium carbonate in a specimen of an impure alkaline substance by titration with standard hydrochloric acid. The reaction takes place in accordance with the equation

$$Na_2CO_3 + 2HCl = 2NaCl + H_2O + CO_2.$$

Is the determination more accurate when carried out with normal acid or with deci-normal acid?

Suppose that 0·5 gramme of the substance is weighed out and titrated with normal acid, and that 4·0 cubic centimetres of acid are required. The weight of anhydrous sodium carbonate is equal to $4 \times 0·053$ grammes, and the percentage of it is 42·4.

If in the determination an error of one-tenth of a cubic centimetre in the burette reading is made, that is, if 4·1 c.c. are used
instead of 4·0 c.c., the weight of sodium carbonate is 4·1 × 0·053
grammes, and the percentage of it is 43·4.

If instead of weighing out 0·5 gramme for the determination,
5 grammes are taken and titrated with normal acid, 40 c.c. of acid
will be required, corresponding to a weight of 40 × 0·053 grammes
of sodium carbonate, or a percentage of 42·4.

Suppose that in the determination the reading of the burette
is 40·1 c.c. instead of 40·0 c.c., the error introduced is only one
part in 400, with the result that the percentage of sodium carbonate
will be 42·5.

The result of the example which we are discussing is to show
the advantage of adjusting matters so that a relatively large
volume of liquid is run out from the burette. An error of one-
tenth of a cubic centimetre gives rise to a smaller error the greater
the volume of the liquid run out.

If we modify the procedure in the above experiment by titrating
the alkaline carbonate with deci-normal hydrochloric acid instead
of the normal acid, and take 0·5 gramme of the substance, the
titration with deci-normal hydrochloric acid will require 40 c.c.,
corresponding to 40 × 0·0053 grammes of sodium carbonate, that
is to a percentage of 42·4. If an error of 0·1 c.c. is made, that
is, if the burette reading is 40·1 c.c. instead of 40·0 c.c., the error in
the final result will lead to a percentage of sodium carbonate equal
to 42·5, or approximately to an error of one part in four hundred.

It is clear that we have realized the same degree of accuracy
by working with deci-normal acid and taking the smaller quantity
of substance as we have realized by employing normal acid with
ten times the quantity of substance. When economy of material
is a consideration, the employment of the more dilute acid is
therefore to be recommended.

The advantage of working with a deci-normal standard solution
in preference to one of normal strength is however obviously only
secured if the change of colour denoting the completion of the
reaction, or the end-point as it is called, is defined with equally
great precision in dilute solution to what it is in a more concentrated solution. This is by no means always the case; and in the

particular example which has been quoted, where the end point is usually defined by the addition of a drop of methyl orange (Chapter VIII), the change of colour is somewhat more sharply defined in solutions of normal strength than in solutions of deci-normal strength. For this reason, it is inadvisable to frame any hard and fast rules for the employment of more dilute rather than of more concentrated standard solutions; all that can be said is that the conditions should be adjusted so that as large a volume of liquid is allowed to flow from the burette as possible; in no case should less than 10 c.c. be run in.

In discussing the accuracy of volumetric determinations, particularly as regards the calculation of results from experiments, it is of course essential to know how much reliance is to be placed upon the measuring vessels. Ordinary pipettes and burettes are usually to be relied upon to an accuracy of the order of one-half per cent. If greater accuracy than this is required it is necessary to calibrate the measuring vessels by filling with water and determining the weight of water which they contain or deliver. If we are dealing with ordinary apparatus which has not been specially calibrated, it is not only mere waste of time to continue the arithmetical work beyond a certain point, but the result has no definite meaning. The process of calculation should be carried only a *little* way beyond the degree of accuracy of the experimental work, in order to ensure that no error is introduced into the final result by the process of calculation.

In connexion with the accuracy of measuring vessels, it is important to bear in mind that the process of weighing out the substance for the standard solution need not be carried out with greater precision than corresponds to the accuracy of the vessels which are to be employed to measure the standard solution. As a matter of fact the process of weighing should be carried out a *little* more accurately than the accuracy of the measuring vessels in order to ensure that the maximum accuracy of the standard solution is realized.

In carrying out volumetric determinations, it is sometimes asked by beginners whether to run the standard solution from the burette into the solution the strength of which is being determined, or to reverse the process, that is, to measure out a known volume of the standard solution and titrate with the solution which is under investigation. It is obvious that if both measuring vessels are equally accurate, and if the change of colour denoting the end-point of the reaction is defined with equal precision in either direction, the result should be identical whichever way the titration is carried out. From that point of view it is immaterial whether the standard solution is used in the pipette or in the burette; nevertheless it will in most cases be found more convenient to run the standard solution from the burette into the unknown solution. In cases, however, in which the colour change is defined with greater precision by reversing the usual procedure, the advantage of greater accuracy should be secured. This is particularly to be observed in cases in which the strength of the standard solution has been determined under special conditions which require to be uniformly satisfied in order to obtain consistent results, as is the case with some volumetric processes which are purely empirical.

It cannot be too strongly impressed upon students that in volumetric analysis, as in all other quantitative experimental work, it is a fatal mistake to be satisfied with a single observation. The student should make it a rule for himself never to calculate the result of a determination from a single titration. Two or three concordant titrations should always be obtained before proceeding to the calculation, which is to be carried out with "approximate" atomic weights. "Accurate" atomic weights are only to be used when carefully calibrated measuring vessels are employed in the experiments.

Preparation and storage of standard solutions

Solutions of known strength may readily be prepared by dissolving the weighed quantity of the pure substance in water and diluting to the necessary volume. When large quantities of standard solution are required, and when the purity of the dissolved substance is open to doubt, solutions of approximate strength may be prepared and the exact strength determined by titration with a solution of accurately known strength prepared from the purest materials. In all cases of making up large quantities of standard solutions from substances which are only approximately pure, it is advisable to take a slightly greater quantity of substance than the theoretical, in order that a solution of the required strength may be obtained by dilution after the exact strength has been determined by titration. If solutions of normal or sub-normal strength are required and a solution of a strength approximating closely to the exact value has been prepared, it is often more

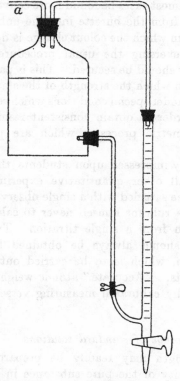
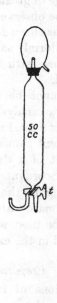

Fig. 1. Reservoir in direct communication with burette. The tube *a* may be placed in direct connexion with an apparatus for generating hydrogen.

Fig. 2. Automatic delivery pipette. The apparatus is filled or emptied by the tap *t*.

convenient to determine by titration how far the solution deviates from the strictly normal or sub-normal strength, and to express the deviation in terms of a factor, than to dilute by adding more

water, or to concentrate by evaporation to make the strength of
the solution the exact value. For many purposes, however, it is
more convenient to work with solutions the exact strength of which
is known and to calculate the results of the analyses from the
general equation (p. 4)

$$\frac{v_1 x}{v_2 w} = \frac{\text{equivalent of } A}{\text{equivalent of } B}.$$

The permanence of standard solutions varies greatly with the
nature of the dissolved substance. Some substances, like potassium dichromate, are absolutely permanent in aqueous solution,
so that if precautions are taken to guard against loss of solvent
by evaporation, the strength of the solutions will remain unchanged
for an indefinite period. Other substances, like sodium thiosulphate, are much less stable in solution, so that the standard
solutions should be titrated at frequent intervals in order to verify
their exact strength. Some substances like potassium permanganate are somewhat sensitive to light; solutions of such substances
should therefore be preserved in dark blue bottles and kept in
the dark.

For many purposes, when the standard solutions require to be
protected from the air, as is the case with titanous chloride, it is
convenient to store the solution in a reservoir which is in direct
communication with a burette (Fig. 1) or with an automatic
delivery pipette (Fig. 2).

The thermal expansion of solutions

In the preparation of standard solutions, the influence of
temperature on the volume of the resulting liquid must be carefully
borne in mind. The coefficient of expansion of solutions is by no
means identical with that of pure water; in general, solutions
possess a higher coefficient. Consequently if heat has been employed in dissolving a solid for the preparation of a standard
solution, sufficient time must be allowed for the liquid to acquire
a constant volume before diluting it to a standard volume. The
same remark applies to the case of substances which dissolve in
water with considerable absorption or evolution of heat. The
temperature coefficient of expansion of solutions of moderate

strength is in general considerably greater than that of more dilute solutions. This is an additional reason for preferring solutions of the order of deci-normal strength to those of normal strength.

In connexion with the thermal expansion of liquids, mention may be made of the degree of accuracy with which burettes should be read. The burettes in common use are graduated in tenths of a cubic centimetre, and may be read with accuracy to one-twentieth. Even in the case of burettes which are calibrated accurately, it is mere waste of time to attempt to read the volume of liquid to any greater degree of accuracy than this, unless the temperature of the liquid is maintained within narrow limits.

Reactions in aqueous solution

When a substance is dissolved in water, the properties of the resulting solution indicate that in many cases a change takes place. It is by no means always an easy matter to say whether the change is to be regarded as a physical or a chemical one. In some cases it appears that the substance dissolves without change. That is to say, the properties of the solution depending upon osmotic effects indicate that the substance in solution possesses a molecular weight equal to that which is arrived at by other methods. Alcohol and cane sugar dissolve to form solutions of this kind. Again, hydration may take place. Hydration is a true chemical change, the substance actually uniting with one or more molecules of water when brought in contact with the solvent. Further, hydrolysis may take place. It may be stated that any reaction in which water plays an essential part other than by mere addition is termed hydrolysis. This phenomenon is of exceedingly common occurrence and manifests itself more particularly in connexion with the appearance of acidic or basic properties when certain salts are dissolved in water. It is of great importance in reactions involving neutralization. Lastly, the properties of solutions of that large class of substances known as salts are of such a nature that the theory has been put forward that these substances undergo electrolytic dissociation to a greater or less extent when dissolved in water. Solutions of substances which are electrolytically dissociated are extremely reactive, and the reactions are so rapid as to be practically instantaneous. Such reactions play a very prominent part in volumetric analysis; but for most purposes it is unnecessary to introduce the ionic theory in the discussion of volumetric methods, and it is perhaps preferable to avoid doing so. The theory of indicators (Chapter IX) is however so very greatly facilitated by the hypothesis of ionization that no apology is necessary for making free use of it in that connexion.

CHAPTER II

DETERMINATIONS WITH STANDARD
POTASSIUM PERMANGANATE

Potassium permanganate, $KMnO_4$, is a dark crystalline solid, which dissolves in water forming a beautiful deep purple solution. This solution possesses very powerful oxidizing properties. Its maximum oxidizing action is obtained when it is employed in presence of free acid. For this purpose sulphuric acid is employed. Nitric acid being itself possessed of oxidizing properties is inadmissible for volumetric work. Hydrochloric acid has been employed in certain cases, but the use of this acid is not to be recommended, for reasons which we shall explain later. In order to understand the oxidizing power of potassium permanganate in presence of sulphuric acid, it is necessary to bear in mind that the essential action consists in the decomposition of the permanganate in accordance with the equation

$$2KMnO_4 + 3H_2SO_4 = K_2SO_4 + 2MnSO_4 + 3H_2O + 5O.$$

The five atoms of oxygen are not liberated in the gaseous state but are given up to any oxidizable substance which may be present in the solution. Thus oxalic acid is converted into carbon dioxide and water

$$5H_2C_2O_4 + 5O = 10CO_2 + 5H_2O.$$

Again, we may represent the oxidation of a ferrous salt as consisting in the conversion of ferrous oxide FeO into ferric oxide Fe_2O_3 as follows

$$10FeO + 5O = 5Fe_2O_3.$$

The complete equation for the oxidation of oxalic acid is therefore

$$2KMnO_4 + 3H_2SO_4 + 5H_2C_2O_4$$
$$= K_2SO_4 + 2MnSO_4 + 8H_2O + 10CO_2.$$

Again, the complete equation representing the conversion of a ferrous salt into a ferric salt may be written

$$2KMnO_4 + 8H_2SO_4 + 10FeSO_4$$
$$= K_2SO_4 + 2MnSO_4 + 8H_2O + 5Fe_2(SO_4)_3.$$

Many other substances may be oxidized quantitatively by potassium permanganate in presence of sulphuric acid. We shall discuss these later.

We defined a normal solution of a volumetric reagent as a solution containing in one litre that fraction of the molecular weight which corresponds to one gramme of available hydrogen. Now in the case of potassium permanganate, since two molecules of $KMnO_4$ give rise to five atoms of available oxygen, and since five atoms of oxygen are equivalent to ten atoms of hydrogen, the normal solution of this substance must contain one-fifth of the molecular weight in grammes in one litre. In other words, the normal solution should contain 31·6 grammes of the salt per litre. It is however more usual to employ this reagent of deci-normal strength, the solution containing therefore 3·16 grammes of the substance dissolved in one litre.

Potassium permanganate is now readily obtainable of a very fair degree of purity. A solution of approximately deci-normal strength may therefore be readily prepared by dissolving the requisite quantity of the substance in water and diluting the solution to the calculated volume. It is necessary to determine the exact strength of the solution; and this may be done in either of the two following ways.

(a) *By ferrous ammonium sulphate*, $FeSO_4(NH_4)_2SO_46H_2O$. This double salt may be prepared in a high degree of purity by recrystallization of the commercial product. Or the salt may be prepared by dissolving in separate small quantities of hot water amounts of ferrous sulphate, $FeSO_47H_2O$, and of ammonium sulphate, $(NH_4)_2SO_4$, in the proportions of their respective molecular weights. It is advisable to add a few drops of dilute sulphuric acid to the iron solution to prevent the formation of basic salt. On mixing the two solutions and cooling, the double salt, being less soluble in water than the single salts, separates in the form of small crystals. These crystals are washed with cold water, recrystallized, and finally dried by pressing between filter paper. A suitable quantity, say 4 or 5 grammes, is then weighed out carefully, dissolved in water, and made up to some definite volume. Aliquot portions are then withdrawn, acidified with dilute sulphuric acid, and the solutions titrated with the solution of potassium permanganate. The permanganate is added from a burette provided with a glass tap: a rubber joint must not be used as it would reduce the permanganate. No indicator is added to the liquid, since the slightest excess of permanganate imparts a permanent pink colour to the solution. Several titrations must be made, and they must agree with one another to one-half per cent. The double iron salt contains almost exactly one-seventh of its weight of iron, as may be seen by calculating its percentage composition from its molecular formula.

If v_1 c.c. of the ferrous solution require for complete oxidation v_2 c.c. of $KMnO_4$, the strength of the solution of potassium permanganate is given by the equation

$$\frac{v_1 \times \text{weight of iron in 1 c.c.}}{v_2 x} = \frac{\text{equivalent of iron}}{\text{equivalent of permanganate}} = \frac{10 \times 56}{2 \times 158},$$

where x is the weight of the permanganate per c.c.

(b) *By oxalic acid*. As we have mentioned already, oxalic acid is oxidized quantitatively to carbon dioxide and water by potassium permanganate in presence of dilute sulphuric acid. The recrystallized acid containing two molecules of water of crystallization serves admirably therefore for the standardization of permanganate. Some operators prefer to use sodium oxalate, $Na_2C_2O_4$, which crystallizes anhydrous, as there seems to be a slight doubt about the constancy of the water of crystallization of the free acid. In either case the method of procedure is the same. A standard solution of oxalic acid or of sodium oxalate is prepared by dissolving a weighed quantity of the substance in water and diluting the solution to the required volume. Aliquot portions are then withdrawn for the titration, acidified with dilute sulphuric acid, and warmed to about 60° or 70° C. and the permanganate solution added slowly until a faint permanent pink colour remains in the liquid. The reason for warming the solutions previous to titration is that the velocity of the reaction between oxalic acid and potassium permanganate is not great at the ordinary temperature, but as is usually the case, the reaction velocity is enormously increased by an elevation of temperature. Once the reaction has started, the manganous sulphate, which is formed as a by-product of the reaction, acts as a catalyst on the oxidation of the oxalic acid. Oxalic acid may indeed be titrated at the ordinary temperature if a quantity of manganous sulphate is added at the beginning, but the process is in any case considerably slower than if the solution be warmed. The reaction between oxalic acid and potassium permanganate in presence of sulphuric acid is an interesting example of autocatalysis, *i.e.* of catalysis of the main reaction by one of the products of the reaction.

The calculation of the strength of the solution of potassium permanganate is made from the equation

$$\frac{v_1 \times (\text{weight of oxalic acid per c.c.})}{v_2 x}$$
$$= \frac{\text{equivalent of oxalic acid}}{\text{equivalent of permanganate}} = \frac{5 \times 126}{2 \times 158}.$$

In titrating oxalic acid or a ferrous salt by permanganate it is important not to add the reagent too rapidly, otherwise a brown precipitate of hydrated higher oxides of manganese may be produced, and it is exceedingly difficult to remove this precipitate afterwards. All titrations in which a precipitate appears should be rejected.

A standard solution of potassium permanganate will retain its strength for a long period, especially if preserved from light by storing it in dark blue bottles.

Determination of ferrous and ferric iron by potassium permanganate

The determination of ferrous salts has virtually been described in discussing the use of ferrous ammonium sulphate as a reagent for standardizing permanganate, and no further discussion is necessary. Ferric salts must first be reduced to the ferrous condition, and the resulting solution titrated by permanganate in the ordinary way. Of the various substances employed as reducing agents, the following are the most generally used: zinc and sulphuric acid, hydrogen sulphide, and sulphur dioxide.

The procedure for reducing ferric solutions by means of zinc and sulphuric acid is as follows. Aliquot portions of the iron solution are placed in separate flasks, a moderate amount of granulated zinc and dilute sulphuric acid added to each and the contents warmed. From time to time a drop of the liquid undergoing reduction is removed by means of a glass rod and brought in contact with a drop of ammonium thiocyanate on a white tile. As long as any ferric iron remains, a dark red colour is produced due to the formation of ferric thiocyanate, but when the iron is completely converted to the ferrous condition, the drops on the spot plate remain colourless when tested with the iron solution. The solution of ferrous sulphate is next separated from undissolved zinc by filtration through glass wool. Water is next added to the flask, allowed to flow through the glass wool, and the washings added to the reduced filtrate, which is titrated with potassium permanganate in the usual way.

The reduction of ferric salts to the ferrous condition by means of zinc and dilute sulphuric acid is frequently regarded as taking place in consequence of the so-called nascent hydrogen generated by the action of the metal on the acid, since it is impossible to effect the reduction of the ferric salt by simply bubbling hydrogen through the solution. It must be pointed out, however, that no such explanation can be regarded as satisfactory, whatever views are entertained with regard to the nascent state. If there be any truth in the idea of nascent hydrogen, it is obvious that this substance should possess the same properties whatever be the source from which it is derived. Now it has been shown that solutions of vanadium corresponding to the pentoxide V_2O_5 are reduced to the dioxide V_2O_2 by zinc and hydrochloric acid, but when magnesium is employed instead of zinc, then salts of vanadium corresponding to the trioxide V_2O_3 are obtained. This behaviour of vanadium salts is a striking example of the futility of the idea of nascent hydrogen as a definite reducing agent independent of the source from which it is obtained.

In reducing a solution of ferric sulphate to the ferrous condition by means of zinc and sulphuric acid, it is advisable in all cases to follow the method of procedure already described, viz., to add a considerable excess of granulated zinc and to separate the excess of the metal by filtration after the reduction of the iron solution is complete, rather than to add only a small quantity of zinc and allow the action to continue until all the metal has dissolved, because the reduction takes place essentially at the surface of the zinc, and the process is all the more rapid the greater the surface of metal exposed to the solution. The reduction of ferric salts by means of zinc yields excellent results. It is perhaps scarcely necessary to point out that the zinc used must be free from iron and other impurities such as arsenic.

If sulphuretted hydrogen be selected as the reducing agent, aliquot portions of the iron solution are diluted with water, acidified with a little dilute sulphuric acid, and saturated with the gas

$$Fe_2(SO_4)_3 + H_2S = 2FeSO_4 + H_2SO_4 + S.$$

When it has been ascertained by the thiocyanate test that reduction is complete, the solution is boiled to expel the excess of hydrogen sulphide, and to granulate the precipitate of sulphur. The sulphur is then filtered off, washed with water, and the filtrate and washings titrated with permanganate.

For many purposes the reduction of solutions of ferric iron may be effected in a very satisfactory manner by the use of sulphur dioxide. The gas may be generated *in situ* by adding a few crystals of sodium sulphite to the measured quantity of the solution of the ferric salt, and acidifying with dilute sulphuric acid; or the solution may be saturated directly with sulphur dioxide by bubbling the gas from one of the syphons of the liquid. In either case it is important to have the solution dilute and to avoid having too much free acid present, otherwise the reduction does not proceed very easily

$$Fe_2(SO_4)_3 + SO_2 + 2H_2O = 2FeSO_4 + 2H_2SO_4.$$

The solution is now heated to boiling and drops are brought in contact with drops of ammonium thiocyanate in order to observe the progress of the reaction. The boiling is continued until the sulphur dioxide is completely expelled. The solution is then cooled and titrated with permanganate.

It is to be observed that while all three reduction methods give satisfactory results as far as the actual reduction of a ferric solution is concerned, circumstances may arise when one or other method is to be preferred. For example, if for any reason it is essential to avoid the introduction of a foreign metal into the solution, then one would make use of hydrogen sulphide or sulphur dioxide as the reducing agent.

Determination of ferrous and ferric iron in a mixture

Measured quantities of the solution are acidified with dilute sulphuric acid, and then titrated with potassium permanganate in the usual way. If n_1 c.c. of permanganate are required, this quantity represents the amount of ferrous iron in the quantity of the mixture taken. The same quantities of the mixture are again measured out and reduced by any of the methods already described. These solutions are then titrated with permanganate. This second titration, n_2 c.c., represents the total iron in the measured quantity of the solution. The ferric iron is therefore represented by $(n_2 - n_1)$ c.c. of potassium permanganate.

Determination of hydrogen peroxide in aqueous solution

Since hydrogen peroxide and potassium permanganate react in presence of sulphuric acid in the manner represented by the equation

$$5H_2O_2 + 2KMnO_4 + 3H_2SO_4 = K_2SO_4 + 2MnSO_4 + 8H_2O + 5O_2,$$

the former substance may be very easily determined. But since commercial hydrogen peroxide frequently contains organic matter to act as a preservative to prevent spontaneous decomposition, this organic matter undergoes oxidation by permanganate, and the method therefore is liable to give high results. Consequently this method is not to be recommended for the determination of this substance. A better method of determining hydrogen peroxide is described in Chapter v.

Determination of ferrocyanides

Potassium ferrocyanide is very easily oxidized to potassium ferricyanide by means of potassium permanganate in presence of dilute sulphuric acid. The reaction takes place in accordance with the equation

$$10K_4FeC_6N_6 + 8H_2SO_4 + 2KMnO_4$$
$$= 10K_3FeC_6N_6 + 6K_2SO_4 + 2MnSO_4 + 8H_2O.$$

In presence of dilute sulphuric acid, the ferrocyanide solution possesses a greenish yellow colour. The ferricyanide on the other hand is possessed of a golden yellow colour. In this reaction the end-point to be aimed at is not the faint pink colour to which the operator is familiar in permanganate work, but the point at which the solution changes from greenish yellow to a rich golden brown. A little practice will soon enable the experimenter to determine the point without difficulty.

Determination of oxidizing agents by means of potassium permanganate

Many oxidizing agents may be determined in a very satisfactory manner by allowing them to react with a known excess of ferrous sulphate, and then determining by titration with potassium permanganate the amount of ferrous iron remaining. The difference between the amount of ferrous iron originally taken and that found by the permanganate titration represents the oxidizing power of the substance which is being determined. Persulphates, for example, may be determined in this way.

Use of potassium permanganate in presence of hydrochloric acid

The simplicity and accuracy of the use of permanganate has led many chemists to attempt to employ it as a volumetric process in presence of hydrochloric acid instead of sulphuric acid in cases where the use of the former acid is necessary. But, unless certain precautions are taken, the titration of ferrous iron by means of potassium permanganate in presence of hydrochloric acid gives rise to inaccurate results. It has been stated by some chemists that hydrochloric acid and potassium permanganate react together with evolution of chlorine according to the equation

$$2KMnO_4 + 16HCl = 2KCl + 8H_2O + 2MnCl_2 + 5Cl_2,$$

and that complications are thereby introduced. It is, however, certain that the formation of chlorine does not take place in *dilute* solutions in the simple manner represented by this equation, since it is impossible to demonstrate the formation of chlorine by merely bringing together *dilute* solutions of hydrochloric acid and potassium permanganate. But if dilute solutions of hydrochloric acid and potassium permanganate are brought together in presence of a ferrous salt, then chlorine is formed in sufficiently great amount to be easily recognizable by its odour; in other words, the evolution of chlorine is in some manner connected with the presence of iron. It has further been shown by Fresenius that if one takes a given volume of a hydrochloric acid solution of a ferrous salt, and titrates it with potassium permanganate, and then adds to the titrated solution an equal volume of the original iron solution, and again titrates, the volume of permanganate required to complete the reaction is greater in the first case than in the second. If the experiment be repeated a third and a fourth time in the manner described, it will be found that the volumes of permanganate will become still smaller, and then remain constant. When constant volumes of permanganate are obtained, it will be found that those volumes are an approximately correct measure of the amount of ferrous iron which is undergoing oxidation. It has also been found that approximately correct results are obtained if a moderate quantity of manganous sulphate be added to a hydrochloric acid solution previous to titration with potassium permanganate.

We have now to ask ourselves two questions; first, in what manner does iron act in causing the formation of chlorine from a mixture of dilute solutions of hydrochloric acid and potassium permanganate, and, secondly, how does the addition of manganous sulphate to the solution prevent the formation of chlorine? It would scarcely be within the scope of the present work to enter into an elaborate discussion of the various explanations which have been offered for these curious phenomena, but it may be stated that the theory which has perhaps attracted most attention is that developed by Manchot and others, according to which the iron while being converted from the condition corresponding to ferrous oxide, FeO, absorbs oxygen so rapidly from the permanganate that a higher oxide of iron than ferric oxide, Fe_2O_3, is formed, perhaps Fe_2O_5. This higher oxide then decomposes into ferric oxide and oxygen, which reacts with the hydrochloric acid with development of chlorine. If, however, a manganous salt is present in solution, this salt is capable of acting as a carrier of oxygen, the manganous salt being oxidized by the iron peroxide to a higher oxide of manganese, and this higher oxide of manganese effects the oxidation of more ferrous iron.

The oxidation of a ferrous salt by acidified potassium permanganate may be represented in the following manner

$$Mn_2O_7 + 2FeO = 2MnO_2 + Fe_2O_5 \quad \ldots\ldots\ldots (1),$$
$$Fe_2O_5 + 4FeO = 3Fe_2O_3 \quad \ldots\ldots\ldots\ldots (2).$$

The manganese peroxide formed according to equation (1) also reacts with more ferrous oxide, resulting in the formation of ferric oxide

$$2MnO_2 + 4FeO = 2Fe_2O_3 + 2MnO \quad \ldots\ldots\ldots (3),$$

so that the familiar result of ten molecules of ferrous salt becoming oxidized by two molecules of permanganate into five molecules of ferric salt is obtained. Reaction (2), according to Manchot (*Annalen der Chem.* **325**, 1902, p. 105), takes place in sulphuric acid solution with such great rapidity that no free oxygen is liberated, and consequently the results are accurate.

In presence of hydrochloric acid, however, the iron peroxide reacts with the hydrochloric acid resulting in the formation of chlorine, according to the equation

$$Fe_2O_5 + 10HCl = 2FeCl_3 + 5H_2O + 2Cl_2 \quad \ldots\ldots (4).$$

But, if a manganous salt is present in sufficient quantity, the production of chlorine is prevented in two ways, first, because the manganous salt reacts with the permanganate forming hydrated manganese peroxide according to the equation

$$Mn_2O_7 + 3MnO = 5MnO_2 \quad \ldots\ldots\ldots\ldots (5),$$

and, secondly, because the manganous oxide reacts with the iron peroxide resulting in the formation of ferric oxide and manganese peroxide

$$Fe_2O_5 + 2MnO = 2MnO_2 + Fe_2O_3 \quad \ldots\ldots\ldots (6).$$

This explanation of the formation of chlorine during the titration of a hydrochloric acid solution by means of potassium permanganate, and the prevention of the evolution of chlorine by the addition of a sufficient quantity of a manganous salt, seems to accord fairly well with the observed facts; but, nevertheless, the titration of iron by potassium permanganate in presence of hydrochloric acid is not to be recommended, as undoubted irregularities have been found to occur even in presence of excess of a manganous salt. The method is at best an empirical one. In particular, it has been shown by Birch (*Chem. News*, 1909, pp. 61 and 73) that the method of conducting the determination as recommended by Fresenius does not give accurate results. If it is necessary to determine iron in presence of hydrochloric acid, the best plan is to employ potassium dichromate as the oxidizing agent.

Determination of metals which form insoluble oxalates

Inasmuch as oxalic acid may be determined by permanganate with great ease and accuracy, metals which are quantitatively precipitated as oxalates may be estimated by decomposition of the oxalate by means of dilute sulphuric acid, and titration of the resulting oxalic acid by standard potassium permanganate. As an example we shall describe the procedure for the determination of calcium in Iceland Spar ($CaCO_3$).

A suitable quantity of Iceland Spar is weighed out and dissolved in excess of dilute hydrochloric acid, care being taken to avoid loss of liquid by spirting. The solution is next made alkaline with ammonia, a small quantity of ammonium chloride added, and the calcium precipitated by addition of excess of ammonium oxalate. The precipitate is next collected, washed free of the excess of the precipitant, and then decomposed with dilute sulphuric acid. The solution is next made up to a known volume with distilled water, and aliquot portions are titrated with potassium permanganate.

Since two molecules of $KMnO_4$ are equivalent to five molecules of oxalic acid, this quantity of permanganate is equivalent to five atoms of calcium.

It is clear that the process is applicable to any metal which forms an insoluble oxalate. The process may be modified by precipitating the metal with a known excess of a soluble oxalate, and determining by means of permanganate the quantity of oxalate which remains unprecipitated. This latter method, however, does not possess any advantage over that already described.

Potassium permanganate may be employed for the volumetric determination of certain of the rarer elements such as exist in more than one state of oxidation, but we must refer the reader to special treatises for a detailed description of the procedure.

Determination of copper by potassium permanganate

If to a solution of copper a metal be added which is more electro-positive than copper, such as zinc, substitution takes place, the zinc going into solution and the copper being precipitated

$$Zn + CuSO_4 = ZnSO_4 + Cu.$$

The copper, which is precipitated in a finely divided form, is filtered off, washed with distilled water, and dissolved in a mixture of ferric sulphate and sulphuric acid. The metal dissolves and at the same time the ferric sulphate is reduced quantitatively to ferrous sulphate

$$Cu + Fe_2(SO_4)_3 = CuSO_4 + 2FeSO_4.$$

The ferrous sulphate which is formed is then determined volumetrically with standard potassium permanganate in the usual way.

Since one atom of copper liberates two molecules of ferrous sulphate, and since ten molecules of ferrous sulphate are oxidized quantitatively to ferric sulphate by two molecules of potassium permanganate, it is clear that five atoms of copper are equivalent to two molecules of potassium permanganate, or adopting atomic weights, $5 \times 63 \cdot 6$ grammes of copper are equivalent to 2×158 grammes of potassium permanganate.

In carrying out the determination of copper by this method, it is necessary to have the metal present in the form of the sulphate. Nitric acid and certain heavy metals, especially bismuth and lead, must be absent. Either the zinc may be employed in the form of sticks which are well washed when the whole of the copper has been precipitated from the solution, or the granulated metal may be employed, and sufficient sulphuric acid added to complete the solution of the zinc after precipitation of the copper. In order to test that precipitation of the copper from the solution is complete, the best plan is to bring a drop of the liquid in contact with a drop of potassium ferrocyanide solution on a white tile when no brown colour should appear. This method of determining copper yields excellent results, the chief disadvantage of the method being the fact that metallic copper is dissolved very slowly by an acidified solution of a ferric salt.

Determination of copper by reduction to cuprous oxide and subsequent determination by ferric sulphate and permanganate

If excess of sodium hydroxide be added to a solution of a copper salt, a blue precipitate of hydrated copper oxide is obtained. This precipitate is readily soluble in solutions of certain organic compounds containing hydroxyl groups, such as tartaric acid. From a solution of cupric oxide in an alkaline tartrate, reducing agents such as glucose will readily precipitate cuprous oxide, Cu_2O, as a red powder on warming. The precipitation is quantitative, and the metal is readily determined by separating the cuprous oxide and dissolving it in a solution of a mixture of ferric sulphate and sulphuric acid, when reduction of the iron salt occurs in accordance with the equation

$$Cu_2O + H_2SO_4 + Fe_2(SO_4)_3 = 2CuSO_4 + 2FeSO_4.$$

The resulting ferrous sulphate is then determined by titration with potassium permanganate in the ordinary manner.

In carrying out the determination of copper by this method, the liquid is made strongly alkaline with caustic soda and excess of a solution of potassium sodium tartrate, $KNaC_4H_4O_6$, is added. An aqueous solution of grape sugar is then added in moderate excess, and the liquid kept at a temperature of about 70° C. until the precipitate of cuprous oxide assumes a bright red colour, and the supernatant liquid is of a brownish yellow colour due to the action of the alkali on the sugar. The precipitate is then filtered off, washed thoroughly with water, and dissolved in a solution of ferric sulphate acidified with sulphuric acid. The amount of copper is then determined by the permanganate titration of the ferrous iron, and it will be clear from the above equation that one atom of copper is equivalent to one atom of iron. This method gives very good results and is preferable to that previously described, as cuprous oxide dissolves quickly in acid solutions.

Determination of metals which form insoluble sulphides

Certain metals which are precipitated as sparingly soluble sulphides may be determined by allowing the separated insoluble sulphide to react with a known excess of ferric sulphate, resulting in the formation of an equivalent amount of ferrous sulphate by reduction which is in its turn estimated by titration with standard potassium permanganate. As an example we may refer to the determination of cadmium.

The measured quantity of solution of the cadmium salt which must not contain too much free acid is precipitated by passing sulphuretted hydrogen through the hot liquid. The reaction is represented by the equation

$$CdCl_2 + H_2S = CdS + 2HCl.$$

The precipitated cadmium sulphide is separated by filtration, and the filtrate again saturated with the gas in order to precipitate any cadmium which may remain in solution. This procedure is absolutely necessary since the reaction between cadmium chloride and hydrogen sulphide is reversible. The collected precipitates are digested with a moderate excess of ferric sulphate solution when reduction takes place in accordance with the equation

$$CdS + Fe_2(SO_4)_3 = CdSO_4 + 2FeSO_4 + S.$$

The sulphur is separated by filtration, and the filtrate and washings acidified with sulphuric acid, and titrated with permanganate.

Zinc is a metal which may be determined in this manner, but the precipitation of zinc sulphide must be effected in ammoniacal solution since it is readily soluble in acids. Many chemists do not trouble to remove the sulphur and filter paper before titrating with permanganate; as it has been shown that the reducing effect upon permanganate is practically *nil* if the solution is cold and very dilute. But it is certainly safer to remove the sulphur especially if it is in a finely divided condition.

The valuation of manganese dioxide

Manganese dioxide, MnO_2, is a substance which is very frequently contaminated with impurities. It is important, therefore, to have a reliable method of determining the percentage of MnO_2 which a sample of the commercial substance contains. One method of determining the percentage of MnO_2 in pyrolusite which is frequently employed consists in treating the substance with a measured excess of standard oxalic acid in presence of dilute sulphuric acid when the manganese dioxide oxidizes the oxalic acid to carbon dioxide and water according to the equation

$$MnO_2 + H_2C_2O_4 + H_2SO_4 = MnSO_4 + 2H_2O + 2CO_2.$$

The excess of oxalic acid remaining over is then determined by titration with standard potassium permanganate in the ordinary way.

In carrying out a determination of manganese dioxide by this method, a suitable quantity, say 0·5 gramme of the sample, is weighed out and digested with a measured quantity of oxalic acid which must be in considerable excess of the amount which is theoretically required to react with the quantity of manganese dioxide taken, along with a sufficient quantity of dilute sulphuric acid. The mixture is heated to a temperature in the neighbour-hood of 70° C., and the action allowed to continue until there is no black residue remaining undissolved. The presence of a white residue does not interfere as it probably consists of insoluble siliceous matter. The liquid may now be directly titrated with permanganate, but it is perhaps preferable to make up the solution to a known volume after filtration from any insoluble residue, and then to withdraw aliquot portions of the solution for titration.

It is to be noted that this procedure is simply a method of determining the amount of available oxygen in the sample, since certain other oxides of manganese are capable of decomposing oxalic acid in presence of sulphuric acid. Thus both manganese sesquioxide, Mn_2O_3, and trimanganic tetroxide, Mn_3O_4, can effect the decomposition, and if either of these oxides be present, the calculation of the percentage of the dioxide in the sample will not really indicate the true amount of this constituent.

Another method of determining the percentage of manganese dioxide in pyrolusite is described on p. 54.

Oxidations by potassium permanganate in neutral and alkaline solutions

Although by far the greater number of determinations which are carried out with potassium permanganate are performed in presence of sulphuric acid, the maximum oxidizing action being obtained under those conditions, there are certain cases more particularly among organic compounds in which it is not possible to effect the determinations in acid solution. Some substances can however be determined in neutral or alkaline (sodium carbonate) solution. In neutral or alkaline solution potassium permanganate possesses three atoms of available oxygen, the decomposition of the molecule taking place in the manner represented by the equation

$$2KMnO_4 = K_2O + 2MnO_2 + 3O.$$

In all such cases, manganese dioxide is precipitated in a hydrated form, and the appearance of this precipitate in the liquid undergoing titration is a disadvantage, as it makes the end-point somewhat difficult to recognize.

Formic acid is a substance which when converted into sodium formate by the addition of excess of sodium carbonate can be determined by potassium permanganate in this way. The oxidation of formic acid takes place in accordance with the equation

$$HCOOH + O = H_2O + CO_2.$$

If n_1 c.c. of the solution of formic acid in presence of excess of sodium carbonate require the addition of n_2 c.c. of potassium permanganate, the weight, x, of formic acid in grammes per c.c. is determined by the equation

$$\frac{n_1 x}{n_2 w} = \frac{3 \times 46}{2 \times 158},$$

where w is the weight of potassium permanganate in grammes per c.c.

Owing to the difficulty of determining the end-point of the reaction in presence of the precipitate of hydrated manganese dioxide, some chemists modify the process by adding a measured volume of potassium permanganate to the hot alkaline formate. The amount of permanganate remaining in excess is then determined by acidifying the solution with dilute sulphuric acid, adding a measured volume of a solution of oxalic acid, and allowing the action to continue until the precipitate has completely dissolved. The excess of oxalic acid is then titrated by potassium permanganate. Then in another experiment, a volume of oxalic acid equal to that previously used is directly titrated with permanganate. The amount of formic acid present in the volume of liquid taken is clearly represented by the difference between the total volume of permanganate taken in the first experiment, and that required for the oxidation of the oxalic acid alone. Although the latter method of procedure is more elaborate, it will in most cases be found probably more satisfactory than the direct method.

Determination of manganese by potassium permanganate

If a solution of potassium permanganate be added to a nearly neutral and hot solution of a manganous salt, a precipitate of hydrated manganese dioxide is obtained by the mutual interaction of the two salts. The reaction may be represented by the equation

$$2KMnO_4 + 2H_2O + 3MnSO_4 = 5MnO_2 + 2H_2SO_4 + K_2SO_4.$$

Potassium permanganate reacts here with three atoms of available oxygen. However, the reaction does not always take place in quite so simple a manner as is represented by the equation, temperature being a factor which requires to be carefully observed. Below 85° C. the reaction does take place according to the equation given, above that temperature free oxygen may be evolved leading to too great a volume of permanganate being required. It has also been found that the action is somewhat more regular in presence of certain salts, zinc salts being specially efficacious. It is perhaps scarcely necessary to point out that other oxidizable substances must be absent.

In carrying out a determination of manganese by this method, the standard permanganate is added from the burette to the hot solution of the manganous salt, the vessel being frequently shaken, until a slight excess of the reagent is indicated by the appearance of a permanent pink colour in the liquid. When the conditions are carefully observed, this method is capable of giving satisfactory results.

Note on the formula of potassium permanganate

Potassium permanganate was formerly and occasionally is still represented by the formula $K_2Mn_2O_8$. This formula clearly represents the fact that the substance possesses five atoms of available oxygen, since the molecule may be regarded as composed of $K_2O + 2MnO + 5O$. On the other hand, although the formula may be regarded as convenient from the point of view of volumetric analysis, there is definite experimental evidence in favour of the simpler formula $KMnO_4$. Ostwald has established the empirical law that the difference between the equivalent conductivities of the sodium salts of acids at dilutions of 32 litres and 1024 litres is a function of the basicity of the acid. The difference is approximately 10 units for a monobasic acid, 20 units for a dibasic acid, 30 units for a tribasic acid and so on. Conductivity determinations have shown that permanganic acid is a monobasic acid, and consequently potassium permanganate is to be represented by the formula $KMnO_4$. It is perhaps unnecessary to point out that such evidence applies only to the molecular condition of the substance in solution; as regards the molecular weight of solid potassium permanganate we possess no definite knowledge whatever.

Determinations of the depression of freezing point in dilute solutions lead to the same conclusion.

CHAPTER III

POTASSIUM DICHROMATE AS A VOLUMETRIC OXIDIZING AGENT

Potassium dichromate $K_2Cr_2O_7$ is an orange coloured crystalline solid possessed of powerful oxidizing properties. In aqueous solution and in presence of free acid it is capable of effecting the oxidation of certain substances which are frequently determined volumetrically by potassium permanganate. Potassium dichromate, however, possesses certain undoubted advantages over permanganate. It may be used in presence either of hydrochloric or of sulphuric acid, and the aqueous solution preserves its strength unchanged for an indefinite period. Further, the solution may be used in a burette with a rubber joint. Potassium dichromate is most commonly employed for the determination of iron by oxidation from the ferrous to the ferric condition.

In presence of hydrochloric or sulphuric acid, one molecule of potassium dichromate gives rise to three atoms of available oxygen

$$4H_2SO_4 + K_2Cr_2O_7 = K_2SO_4 + Cr_2(SO_4)_3 + 4H_2O + 3O,$$

so that the conversion of ferrous chloride into ferric chloride takes place in accordance with the equation

$$K_2Cr_2O_7 + 14HCl + 6FeCl_2 = 2KCl + 2CrCl_3 + 6FeCl_3 + 7H_2O.$$

Now since one molecule of potassium dichromate possesses three atoms of available oxygen, and since three atoms of available oxygen are equivalent to six atoms of available hydrogen, the normal solution of this substance should contain one-sixth of 294·2 grammes dissolved in one litre, or approximately 49 grammes per litre. It is usual, however, to employ this reagent of deci-normal strength, the solution containing therefore 4·9 grammes of the salt in one litre.

Standardization of the solution

Potassium dichromate is nowadays obtainable in a state approximating to chemical purity. There is therefore no difficulty in making up any quantity of the solution by dissolving the weighed quantity of the salt in water and diluting the resulting solution to the necessary volume. It is necessary, however, in all cases, to standardize the solution, and this may be done very satisfactorily by means of ferrous ammonium sulphate.

When an acid solution of a ferrous salt undergoes oxidation by potassium dichromate, the orange colour of the dichromate disappears, and at the same time the solution becomes green in consequence of the presence of the chromic salt which is formed. No precise change of colour occurs at the point at which all the iron is converted into the ferric condition. It is necessary therefore to determine the end-point of the reaction by means of an external indicator. For this purpose potassium ferricyanide is always employed. Ferrous salts react with the ferricyanide with formation of blue ferrous ferricyanide (Turnbull's blue), while ferric salts merely produce a brown colour with potassium ferricyanide. In order to employ this reagent successfully, it is necessary to have the solution freshly prepared and dilute. Drops of freshly prepared and very dilute potassium ferricyanide are placed on a white tile, and drops of the liquid undergoing titration are removed by means of a glass rod, and brought in contact with the ferricyanide. The glass rod is of course cleaned after each test. As long as any ferrous iron is present, the blue colour of ferrous ferricyanide makes its appearance, but when the iron is completely in the ferric condition, only a faint brown colour is formed. The end-point can be determined with great exactness if the ferricyanide solution is prepared properly: a little practice will soon enable the operator to ascertain the strength of solution which gives the best result.

The calculation of the strength of the potassium dichromate

solution from the titration by ferrous ammonium sulphate is made from the following equation

$$\frac{v \times \text{weight of iron per c.c.}}{v' \, x}$$

$$= \frac{\text{equivalent of iron}}{\text{equivalent of potassium dichromate}} = \frac{6 \times 56}{294},$$

where v denotes the volume of the iron solution taken for the titration, v' the volume of potassium dichromate solution required for the complete oxidation of the iron, and x the weight of potassium dichromate in each cubic centimetre of the solution.

Determination of ferric iron by dichromate

As in the case of the determination of ferric iron by potassium permanganate, so in the use of potassium dichromate, it is necessary first to reduce solutions containing iron in the ferric condition to the ferrous state. As far as the processes of reduction by hydrogen sulphide and by sulphur dioxide are concerned, the procedure is exactly as that already described in Chapter II. The use of zinc and sulphuric acid for the reduction of ferric solutions which are subsequently to be titrated by potassium dichromate is not to be recommended, since the zinc in solution reacts with the potassium ferricyanide employed as an indicator with formation of zinc ferricyanide, thereby greatly obscuring the sharpness of the end-point. There is, however, one very good method of reduction available for dichromate titrations, but it should not be employed with permanganate. This consists in the use of stannous chloride as a reducing agent, which reacts with ferric chloride according to the equation

$$SnCl_2 + 2FeCl_3 = 2FeCl_2 + SnCl_4.$$

The excess of the reducing agent is removed from the solution by precipitation with excess of mercuric chloride

$$SnCl_2 + 2HgCl_2 = 2HgCl + SnCl_4.$$

In carrying out the reduction by means of stannous chloride, the measured quantity of the iron solution is diluted somewhat with water, and a moderate quantity of hydrochloric acid added;

stannous chloride solution is then added cautiously to the heated iron solution until the colour vanishes. A skilled experimenter can judge with the eye when reduction is complete, but it is certainly advisable to test in the usual way with ammonium thiocyanate that all the iron is in the ferrous condition. After cooling, the excess of stannous chloride is removed by the addition of a slight excess of mercuric chloride. This method of reduction when carefully carried out gives extremely satisfactory results. It is important to avoid adding too large an excess of stannous chloride to effect the reduction of the iron, as a correspondingly large excess of mercuric chloride must be added afterwards, and while the presence of a moderate amount of mercury salts does not interfere with the titration, a great excess seems to render the results less accurate.

Determination of the reducing power of a solution of stannous chloride

A solution of stannous chloride finds application in a number of volumetric processes as a quantitative reducing agent. Owing to its liability to undergo oxidation, a solution of stannous chloride should be preserved in an atmosphere of hydrogen, or should at least be protected from atmospheric oxygen by storing it in a bottle to which air can gain access only through a strongly alkaline solution of pyrogallol.

The reducing power of such a solution is very easily determined by allowing known volumes of the solution to react with an excess of ferric chloride solution, and then determining by titration with standard potassium dichromate the amount of ferrous iron produced. From the equation

$$SnCl_2 + 2FeCl_3 = SnCl_4 + 2FeCl_2$$

it is clear that 190 parts by weight of stannous chloride react with $2 \times 55 \cdot 85$ parts by weight of metallic iron. If the dichromate titration shows that w grammes of ferrous iron per c.c. are formed after reduction with stannous chloride, this is clearly equivalent to $\dfrac{190w}{2 \times 55 \cdot 85}$ grammes of stannous chloride per c.c. Very frequently however the reducing power of the solution is expressed in terms of metallic iron, *e.g.* 1 c.c. of the stannous chloride is equivalent to *g* grammes of iron.

It is clear from what has been stated that the above process constitutes a method for the determination of tin in the stannous condition, 119 parts by weight of that metal being equivalent to $2 \times 55\cdot85$ parts by weight of iron. The method gives very satisfactory results.

Potassium dichromate is of very great value as a volumetric reagent for the estimation of iron in ores. The procedure to be followed in order to effect the solution of the ore varies slightly with the nature of the other constituents in the ore, and a special treatise should be consulted for details, particularly as regards those ores which are not completely soluble in acids. As has been stated already, the availability of potassium dichromate in presence of hydrochloric acid, as well as the fact that its aqueous solution may be preserved of constant strength for an indefinite length of time, makes it preferable to potassium permanganate for many purposes. On the other hand it possesses the disadvantage of requiring an external indicator, which makes the titrations take a longer time.

All substances which oxidize ferrous salts quantitatively to the ferric condition can be determined by a similar method to that already described in connexion with potassium permanganate, viz., the addition of a known excess of the ferrous salt to the oxidizing agent, and the subsequent titration by dichromate of the amount of ferrous iron remaining in excess.

Standardization of potassium dichromate by means of iron wire

Iron wire was formerly, and still is, largely used to determine the exact strengths of solutions of potassium dichromate and also of potassium permanganate; a solution of ferrous sulphate being prepared by dissolving a known weight of iron wire in dilute sulphuric acid, and making the solution up to some definite volume. It is not difficult to obtain iron wire containing $99\cdot6$ per cent. of pure iron, and for most purposes this degree of purity is quite sufficient; but if the strength of the volumetric oxidizing agent is required to be known with great exactness, pure iron prepared by the electrolysis of ferrous oxalate should be employed. When iron of the order of 99·6 per cent. purity is employed, some operators multiply the weight of iron taken by 0·996 in order to correct for the small amount of impurity present, but it is very doubtful if the correction is a justifiable one, as it has been found that some of the impurities have a reducing action on permanganate.

In effecting the solution of iron wire, it was formerly the custom to take somewhat elaborate precautions to prevent the access of air to the solution in order to avoid oxidation of the solution; but it has gradually become recognized that solutions of ferrous salts are much more stable towards atmospheric oxygen than is commonly supposed. Solutions of ferrous sulphate are certainly much more stable than solutions of ferrous chloride; it is advisable therefore to use sulphuric acid in preference to hydrochloric for dissolving the metal. For most ordinary work, the solution of the metal may be effected very conveniently in a flask which is heated on a hot plate, a small funnel being placed in the neck of the flask during the operation. The contents are then washed carefully into a standard flask and diluted to a known volume. Aliquot portions are then taken for titration by potassium dichromate.

CHAPTER IV

DETERMINATIONS WITH STANDARD IODINE

A solution of iodine in potassium iodide finds frequent application in analysis as a volumetric oxidizing agent. Thus it is capable of converting arsenious and antimonious compounds into the corresponding arsenic and antimonic compounds. If employed in conjunction with a standard solution of sodium thiosulphate, its range of application is capable of being extended considerably. We shall, however, in the present chapter restrict our discussion to the determination of certain reducing agents by standard iodine alone.

Preparation of a deci-normal solution of iodine

Since iodine is a monovalent element and since its atomic weight is 126·92, a normal solution will contain that weight of the element in grammes dissolved in one litre. In practice one never employs a solution of greater strength than deci-normal; the necessary quantity of iodine in a solution of deci-normal strength will therefore be 12·692 grammes per litre or for approximate purposes 12·7 grammes per litre.

Commercial resublimed iodine may contain various impurities including cyanogen iodide. It may be purified to a considerable extent by mixing it with potassium iodide and subliming. It is, however, in general of no advantage to take elaborate precautions to prepare iodine of a high degree of purity; it is better to prepare a solution of approximately deci-normal strength, and then to determine its exact strength in one of the ways about to be described. In weighing iodine, it is to be observed that the volatility of this element may make an exact weighing a difficult matter, unless the substance be weighed in a closed vessel.

The weighed quantity of iodine is dissolved in a solution of potassium iodide. The weight of potassium iodide to be taken should be from one-and-one-half times to twice that of the iodine. The solution is then diluted with distilled water to the necessary volume. Heat should not be employed to accelerate the dissolution of the iodine on account of the volatility of the element. If the solution be preserved in dark well-stoppered bottles, it will retain its strength unaltered for a long period. This reagent must always be employed in a burette fitted with a glass tap, as it makes rubber tubing useless.

An excellent method of standardizing a solution of iodine is by the use of arsenious oxide purified by resublimation. A small quantity of arsenious oxide is heated in a dish covered with another vessel. The resublimed arsenious oxide should be perfectly white in colour. About 5 grammes of the solid are weighed out accurately, dissolved in an aqueous solution containing about four times the weight of pure sodium carbonate, and then diluted to one litre. In this way a standard solution of arsenious oxide in the form of sodium arsenite is obtained.

When iodine is added to a solution of an alkaline arsenite, oxidation of the arsenic takes place in accordance with the equation

$$As_2O_3 + 2I_2 + 2H_2O \rightleftarrows As_2O_5 + 4HI,$$

two molecules of iodine (four atoms) oxidizing one molecule of arsenious to arsenic oxide. It will be observed, from the above equation, that free hydrogen iodide is formed in the reaction. Now hydrogen iodide is a reducing reagent, and the reaction represented above is reversible. Consequently it is necessary to add some substance which will combine with the hydriodic acid to enable the oxidation of the arsenious oxide to proceed to completion. For this purpose it is usual to employ sodium bicarbonate $NaHCO_3$. Caustic alkalis must not be employed, since they react with iodine with formation of iodide and hypoiodite. A moderate excess of a solution of sodium bicarbonate should therefore be added in each titration.

The end-point is observed with great precision by the addition of a small quantity of a solution of starch to the liquid undergoing titration. The smallest quantity of free iodine in excess will then be recognized by the appearance of the well-known blue colour which this element forms with starch. The preparation of the starch solution is a matter of importance. The best plan is to add to about 100 c.c. of boiling water a small quantity of starch ground up with a little water, and to continue the boiling for a few minutes. It is important to avoid taking too much starch; if too much starch be used the resulting solution will become gelatinous. It is advisable though not necessary to filter the starch solution before using it as an indicator. When carefully prepared, the most minute quantity of free iodine dissolved in potassium iodide can be detected by the appearance of a dark blue colour.

If x be the weight of iodine in grammes in each cubic centimetre of the solution, and if n_1 c.c. of the iodine solution require n_2 c.c. of alkaline arsenite solution containing w grammes of As_2O_3 per c.c., the strength of the iodine solution is determined by the equation

$$\frac{n_1 x}{n_2 w} = \frac{\text{equivalent of iodine}}{\text{equivalent of arsenious oxide}} = \frac{127 \times 4}{198}.$$

Other methods of standardizing iodine

A solution of iodine may be standardized in various other ways than that already described. If a standard solution of sodium thiosulphate is available, the strength of an iodine solution may be determined with accuracy by titrating the one solution against the other, when interaction takes place with formation of sodium iodide and tetrathionate, according to the equation

$$I_2 + 2Na_2S_2O_3 = Na_2S_4O_6 + 2NaI.$$

If the exact strength of the thiosulphate solution is not known, but the experimenter be provided with a standard solution of potassium permanganate or dichromate, the strength of the thiosulphate solution may be determined by allowing the solution of potassium permanganate or dichromate acidified with sulphuric acid to react with an excess of potassium iodide, when iodine is liberated quantitatively in accordance with the equations

$$2KMnO_4 + 8H_2SO_4 + 10KI = 6K_2SO_4 + 2MnSO_4 + 8H_2O + 5I_2$$

and

$$K_2Cr_2O_7 + 7H_2SO_4 + 6KI = 4K_2SO_4 + Cr_2(SO_4)_3 + 7H_2O + 3I_2,$$

in the cases of the permanganate and dichromate respectively. The liberated iodine is then titrated with sodium thiosulphate solution, the end-point being determined with starch, and then the iodine solution, the strength of which is required, is determined by titration against the solution of sodium thiosulphate. The experimental procedure in connexion with the determination of iodine by sodium thiosulphate will be described in greater detail in the next chapter.

Determination of antimony by standard iodine

Antimony in the antimonious condition may be determined with accuracy by titration with iodine, the element undergoing oxidation from the antimonious to the antimonic condition according to the equation

$$Sb_2O_3 + 2I_2 + 2H_2O \rightleftharpoons Sb_2O_5 + 4HI.$$

As in the case of the reaction between arsenious oxide and iodine, so in the present case, the reaction is reversible owing to the reducing action of the hydrogen iodide on the antimonic oxide. For this reason it is as before necessary to suppress the reverse reaction by the addition of a sufficient excess of sodium bicarbonate. It is necessary also to add a sufficient quantity of tartaric acid or of Rochelle salt (potassium sodium tartrate $KNaC_4H_4O_6$) to prevent the precipitation of basic salts of the metal as the result of hydrolysis.

If the metal is present in the antimonic condition to start with, it is necessary to reduce it to the antimonious condition before titration with standard iodine. Sulphur dioxide is an excellent reducing agent for this purpose. The reduction is carried out in presence of hydrochloric acid, and the excess of sulphur dioxide removed by prolonged boiling. The solution is then made just alkaline by caustic soda, and a slight excess of tartaric acid added. After addition of excess of sodium bicarbonate, the solution is titrated with standard iodine in the usual manner, using starch as indicator.

Determination of tin

A solution of iodine is capable of oxidizing tin from the stannous condition to the stannic condition quantitatively, one molecule (two atoms) of iodine effecting the conversion of one atom of tin from the stannous to the stannic condition. In carrying out the determination of tin by this method, a sufficient quantity of Rochelle salt must be present to dissolve any stannous oxychloride or other basic salt, and the solution must be made alkaline by sodium bicarbonate to prevent the reducing action of the hydrogen iodide formed in the reaction from reconverting any stannic salt into stannous. The end-point is obtained in the usual manner by the use of starch. This method of determining tin gives results of very fair accuracy; the chief precaution to be observed is to prevent the absorption of oxygen from the air by the stannous salt.

Determination of sulphur dioxide in aqueous solution

An approximate determination of sulphur dioxide in solution may be made by allowing the aqueous solution of the gas to react with iodine in accordance with the equation

$$I_2 + SO_2 + 2H_2O = H_2SO_4 + 2HI.$$

It was shown by Bunsen that the reaction only takes place in accordance with the above equation when the solution of sulphur dioxide is fairly dilute (not greater than 0·04 per cent. of SO_2 by weight). The irregularities which occur with more concentrated solutions appear to be due, in part at any rate, to the fact that under certain conditions the hydrogen iodide reduces some of the sulphur dioxide to free sulphur. It has subsequently been found that oxidation of the sulphur dioxide to sulphuric acid takes place quantitatively in the manner indicated by the above equation if the solution of sulphur dioxide is run into the solution of iodine, not *vice versa*, even when the solution of sulphur dioxide is fairly concentrated. Consequently it is not advisable to attempt the direct titration of this substance by means of standard iodine; it is much better to add the solution of sulphur dioxide to a known excess of iodine and then to determine the excess by means of sodium thiosulphate. It will be seen from the above equation that 127 parts by weight of iodine react with 32 parts by weight of sulphur dioxide.

Determination of sulphuretted hydrogen in aqueous solution

Iodine and sulphuretted hydrogen interact with formation of hydrogen iodide and sulphur in accordance with the equation

$$I_2 + H_2S = 2HI + S.$$

For various reasons the direct titration of aqueous solutions of this gas by means of iodine does not give accurate results. It is stated that greater accuracy may be attained by adding the solution of hydrogen sulphide to a measured excess of iodine, and subsequently determining the excess by sodium thiosulphate as in the case of sulphur dioxide already described.

A method of procedure which yields satisfactory results is to allow the solution of sulphuretted hydrogen to react with excess of a standard solution of arsenious oxide in presence of hydrochloric acid, when arsenious sulphide is precipitated in accordance with the equation

$$As_2O_3 + 3H_2S = As_2S_3 + 3H_2O.$$

The precipitated arsenious sulphide is separated by filtration, and the filtrate and washings titrated by means of standard iodine in the usual manner, starch being employed as indicator. Since three molecules of hydrogen sulphide react with one molecule of arsenious oxide, it is clear that 3×34 parts by weight of sulphuretted hydrogen are equivalent to 198 parts by weight of arsenious oxide. In carrying out the determinations, the best plan is to allow the solution of the gas to flow into a measured excess of standard alkaline arsenite solution. Hydrochloric acid is then added in sufficient excess to react distinctly acid to indicators.

The presence of free acid is necessary to precipitate the arsenic, otherwise a colloidal solution of arsenious sulphide will be obtained. The filtrate and washings are then made alkaline by the addition of excess of sodium bicarbonate, and the determination of the excess of arsenious oxide effected by titration with standard iodine.

CHAPTER V

THE DETERMINATION OF IODINE BY STANDARD SODIUM THIOSULPHATE

In the previous chapter we had occasion to refer to the reaction between iodine and sodium thiosulphate resulting in the formation of sodium iodide and tetrathionate. The importance of this reaction is that it affords a means of determining not only free iodine, but all substances which liberate iodine quantitatively from potassium iodide on acidifying; in other words a large number of oxidizing agents may be determined by standard sodium thiosulphate. The reaction between iodine and sodium thiosulphate is one of the most sensitive and accurate in the whole domain of volumetric analysis, and its range of application is very great.

Sodium thiosulphate crystallizes with five molecules of water, but there seems to be a slight doubt regarding the constancy of the water of crystallization of this substance. The exact strength of the solution should therefore be determined in one of the ways about to be described.

Standardization of an approximately deci-normal solution

An approximately deci-normal solution may be prepared by dissolving one-tenth of the gramme molecular weight of sodium thiosulphate pentahydrate in water and diluting the solution to one litre. If one is in possession of a standard solution of iodine, the determination of the strength of the thiosulphate solution is effected with great ease, the end-point being determined in the usual manner with starch.

From the equation

$$I_2 + 2Na_2S_2O_3 = Na_2S_4O_6 + 2NaI$$

it is clear that 127 parts by weight of iodine react with 158 parts by weight of sodium thiosulphate calculated as anhydrous salt.

The solution of sodium thiosulphate may be standardized in other ways with equally good results. In connexion with the standardization of iodine, it was mentioned that a solution of thiosulphate might be determined with reference to a standard solution of potassium permanganate or dichromate. We shall now describe in somewhat greater detail the procedure to be followed in determining the exact strength of a thiosulphate solution by one or other of these oxidizing agents.

A given volume (say 20 c.c.) of a standard solution of potassium permanganate is measured out, acidified with dilute sulphuric acid and diluted somewhat with water. A few crystals of potassium iodide are now added and the liquid stirred. Iodine is liberated quantitatively in accordance with the equation

$$2KMnO_4 + 8H_2SO_4 + 10KI = 6K_2SO_4 + 2MnSO_4 + 8H_2O + 5I_2.$$

Sodium thiosulphate is then added from a burette when the liberated iodine dissolved in the excess of potassium iodide is gradually removed with formation of sodium iodide and tetrathionate. When only a faint yellow colour of iodine remains, starch solution is added, and the addition of sodium thiosulphate continued until the deep blue liquid becomes perfectly colourless. The end-point is determined with great precision; a single drop of sodium thiosulphate should be sufficient to cause a striking disappearance of colour.

Since according to the preceding equation two molecules of potassium permanganate liberate five molecules of iodine, and since one molecule of iodine reacts with two molecules of sodium thiosulphate, it is clear that two molecules of potassium permanganate are equivalent to ten molecules of sodium thiosulphate. If therefore n_1 c.c. of potassium permanganate containing w grammes of the salt per c.c. require n_2 c.c. of sodium thiosulphate, the weight of sodium thiosulphate x in grammes per cubic centimetre is given by the equation

$$\frac{n_2 x}{n_1 w} = \frac{10 \times 158}{2 \times 158}.$$

If potassium dichromate is employed for standardizing the

thiosulphate solution, the procedure to be followed is precisely the same as that above described. In determining the end-point, it is important to remember that a green chromic salt is formed by the reduction of the dichromate; the completion of the reaction is therefore that point at which the solution changes from the deep blue of starch iodide to pale green. As has been explained in the previous chapter, the liberation of iodine from potassium iodide by means of acidified potassium dichromate takes place in accordance with the equation

$$K_2Cr_2O_7 + 7H_2SO_4 + 6KI = 4K_2SO_4 + Cr_2(SO_4)_3 + 7H_2O + 3I_2.$$

One molecule of potassium dichromate is therefore equivalent to six molecules of sodium thiosulphate. Consequently if n_1 c.c. of potassium dichromate, containing w grammes of the substance dissolved in each cubic centimetre of the solution, require n_2 c.c. of sodium thiosulphate of which the weight of salt x in grammes per cubic centimetre is required, x is found by the solution of the equation

$$\frac{n_2 x}{n_1 w} = \frac{6 \times 158}{294}.$$

Determination of oxidizing agents by potassium iodide and sodium thiosulphate

The examples already given of the use of standard potassium permanganate or dichromate for the determination of the exact strength of a solution of sodium thiosulphate may clearly be applied in the reciprocal way. That is to say, if one is in possession of a standard solution of sodium thiosulphate, one may determine with accuracy the strength of a solution of a dichromate or of a permanganate. The number of substances which may be determined by such a procedure is very great, and we shall discuss some individual examples in the course of the present chapter. There is, however, one general remark which must be made before proceeding further. The velocity of the reaction between potassium iodide and some oxidizing agents in acid solution is frequently not great at ordinary temperatures, that is to say, the liberation of iodine takes place slowly in many cases, particularly towards the end of the reaction, when the mass concentration of the oxidizing agent has fallen to a small value. Consequently in titrating the liberated iodine by means of sodium thiosulphate, the blue colour of starch iodide frequently makes its appearance after the titration is apparently completed. More sodium thiosulphate must then be added until the liberation of iodine no longer takes place.

Determination of hydrogen peroxide

This substance may be determined iodometrically with great accuracy. The measured quantity of the solution of the peroxide is acidified with dilute sulphuric acid, and excess of potassium iodide added. Iodine is liberated quantitatively in accordance with the equation

$$2KI + H_2SO_4 + H_2O_2 = 2H_2O + K_2SO_4 + I_2.$$

The liberated iodine is then determined by means of standard sodium thiosulphate in the usual manner; one molecule of hydrogen peroxide being equivalent to two molecules of sodium thiosulphate.

The chemical dynamics of the reaction between hydriodic acid and hydrogen peroxide has been made the subject of an exhaustive investigation by Harcourt and Esson. These investigators showed that when the active mass of the hydriodic acid was kept approximately constant by the addition of known constant amounts of sodium thiosulphate as soon as free iodine made its appearance, the rate of disappearance of the hydrogen peroxide was at every instant proportional to the amount present, or

$$-\frac{dC}{dt} = kC.$$

It is clear from this equation that theoretically the whole of the hydrogen peroxide will only be decomposed after an infinite time, and as a matter of fact inaccurate results are frequently to be traced to stopping the titration too soon.

In determining hydrogen peroxide by this method, it is important to add a considerable excess of sulphuric acid. The necessity of a large excess does not appear to be very obvious, but there appears to be little doubt that the inaccurate results obtained by some chemists are due to the use of too little acid. A method of determining hydrogen peroxide by means of potassium permanganate was described in Chapter II, but as was there explained the organic preservatives added to this substance are themselves oxidized by potassium permanganate, and consequently the results are frequently too high. No such objection can be raised against the iodometric method, as special experiments have shown that it yields perfectly accurate results in presence of any of the usual preservatives such as glycerol.

Determination of copper

An acidified solution of a cupric salt will liberate iodine quantitatively from potassium iodide with formation of an equivalent amount of cuprous iodide at the same time. Upon this reaction an accurate method of determining copper has been based. It has been shown that the reaction is somewhat irregular in presence of certain mineral acids, but that in presence of acetic acid the following reaction takes place:

$$2Cu(C_2H_3O_2)_2 + 4KI = Cu_2I_2 + 4C_2H_3O_2K + I_2.$$

The free iodine is determined by titration with standard sodium thiosulphate.

If the solution of the cupric salt contains any mineral acid, it is necessary to neutralize the excess of free acid by the addition of sodium carbonate in excess, and to continue the addition of the carbonate till a slight precipitate of basic cupric carbonate is obtained. The solution is then acidified by the cautious addition of acetic acid, a large excess of acetic acid being carefully avoided. A few crystals of potassium iodide are then added, when cuprous iodide is precipitated and an equivalent quantity of free iodine is set free. The liberated iodine is titrated by means of sodium thiosulphate, care being taken that the reaction is completed.

When ordinary precautions are taken, this method of determining copper gives accurate results. It is clear from the above equation that two atoms of copper (metal) liberate one molecule of iodine, which in its turn reacts with two molecules of sodium thiosulphate. The weight of copper (metal) x in each cubic centimetre of the solution is therefore found by solving the equation

$$\frac{n_1\, x}{n_2\, w} = \frac{63 \cdot 6}{158},$$

where n_1 and n_2 denote the volumes of the copper solution and of the sodium thiosulphate respectively taken for the titration.

Determination of the available chlorine in bleaching powder

The precise nature of bleaching powder is still enveloped in mystery despite the large number of investigations which have been made on this substance. As is well known, when slaked lime is allowed to absorb chlorine, a product is obtained which is sometimes represented by the formula $CaOCl_2$, although it is very doubtful if there is any justification for assigning a formula to the substance at all. Some of the chlorine behaves as if present in the form of hypochlorite, and it is upon the amount of chlorine present in this form, or available chlorine as it is technically termed, that the bleaching value of a specimen of bleaching powder depends.

Bunsen has devised an excellent method of determining the percentage of available chlorine in bleaching powder by allowing the substance, acidified by acetic acid, to liberate iodine from excess of potassium iodide. The liberated iodine is then determined by titration by means of standard sodium thiosulphate. Since one molecule of chlorine liberates one molecule of iodine, it is clear that 35·5 parts by weight of chlorine are equivalent to 158 parts by weight of sodium thiosulphate.

In carrying out an estimation of the available chlorine in bleaching powder, a suitable quantity of the sample, say from 10 to 20 grammes, is weighed out, and triturated with a little water in a mortar; the milky liquid is then gradually transferred to a measuring flask and diluted to the necessary volume. The contents of the flask are well shaken, and a measured quantity of the turbid solution withdrawn by means of a pipette. A few crystals of potassium iodide are now added, the liquid diluted with a little water, and acidified with acetic acid. The liberated iodine is then determined by means of sodium thiosulphate, and the percentage of available chlorine in the sample is easily calculated.

The reason for acidifying the solution by means of acetic acid instead of hydrochloric acid is that bleaching powder frequently contains calcium chlorate, with the result that chlorine would be liberated by interaction with the hydrochloric acid added, and high results obtained.

Determination of ferric iron iodometrically

An acidified solution of a ferric salt liberates iodine from potassium iodide with formation of ferrous salt in accordance with the equation

$$2FeCl_3 + 2KI = 2FeCl_2 + 2KCl + I_2.$$

The liberated iodine is then determined by titration with standard sodium thiosulphate. It is perhaps unnecessary to add that the iron solution must contain no nitric acid or other oxidizing agent. The solution is acidified by hydrochloric acid. The liberation of iodine from potassium iodide by the oxidizing action of a ferric salt takes place somewhat slowly, especially towards the end of the reaction, and this circumstance is the chief disadvantage of this method of determining ferric iron directly. A rapid and accurate method for the determination of ferric iron is described in Chapter x.

Determination of other oxidizing agents by means of potassium iodide and thiosulphate

In general it may be stated that any substance which will liberate iodine from potassium iodide quantitatively on acidifying may be determined by the iodometric method, and the few examples which have been given will serve to indicate the wide range of this method. But the possibilities of this method have been by no means exhausted; persulphates, for example, may be determined with satisfactory results. It does not follow, however, that the iodometric method is necessarily the best method to employ when other methods are available; in particular the slowness of the liberation of iodine towards the end of the reaction in certain cases is an undoubted disadvantage. Some substances cannot be determined directly by this method, but may be determined by taking advantage of the fact that when heated with concentrated hydrochloric acid they evolve chlorine; the chlorine may be allowed to liberate iodine from potassium iodide and this liberated iodine determined by sodium thiosulphate.

Determination of oxidizing agents by distillation with hydrochloric acid

Insoluble substances such as manganese dioxide may be determined with accuracy by distillation with excess of concentrated hydrochloric acid, when chlorine is liberated quantitatively. In the case of manganese dioxide, the reaction takes place in accordance with the equation

$$MnO_2 + 4HCl = MnCl_2 + 2H_2O + Cl_2.$$

The chlorine is then passed into an excess of potassium iodide solution, when one molecule of chlorine liberates one molecule of iodine. Since one molecule of iodine reacts with two molecules of sodium thiosulphate, it is clear that one molecule of manganese dioxide is equivalent to two molecules of sodium thiosulphate.

Various forms of apparatus have been devised for carrying out distillation determinations of this kind. If ordinary corks are

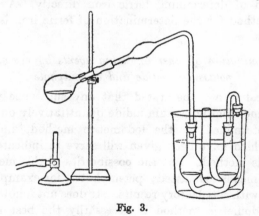

Fig. 3.

employed, they should be soaked in melted paraffin wax before use in order to protect them from the corrosive action of the chlorine, but for the most accurate work it is best to have apparatus constructed with ground glass joints.

For practice in this method a good exercise is the determination of MnO_2 in a specimen of commercial pyrolusite. About

0·5 gramme of the substance is weighed out and introduced into the distillation apparatus, a moderate excess of concentrated hydrochloric acid being added. In order to prevent regurgitation in the event of a fall of the pressure, it is advisable to add a small lump of magnesite to the contents of the distillation flask. This substance dissolves but slowly, generating a steady stream of carbon dioxide, which not only prevents regurgitation but also dilutes the chlorine and renders the absorption of the gas by the potassium iodide less violent. The absorption bulbs should be kept cool by being immersed in water during the experiment. After the reaction is completed, the contents of the absorption tubes are washed out with a little potassium iodide solution, diluted with water to some definite volume; aliquot portions are titrated by means of standard sodium thiosulphate solution, starch being added towards the end of the reaction. From the above equation the percentage of manganese dioxide in the weight of pyrolusite taken may be readily calculated.

The distillation method is clearly applicable to all substances which evolve chlorine when heated with concentrated hydrochloric acid. In general, however, its use is restricted to substances which are insoluble in water, such as lead dioxide and manganese dioxide, since soluble oxidizing agents liberate iodine from potassium iodide directly on the addition of acids.

Use of sodium thiosulphate for residual titration with iodine

In many determinations by means of iodine it is convenient not to determine the end-point of the reaction between the substance which is undergoing oxidation and iodine directly, but to add a measured quantity of standard iodine solution that is known to be in excess of the quantity required, and then to determine the excess of iodine that has been added by residual titration with standard sodium thiosulphate. For example it was pointed out in Chapter IV that the direct titration of sulphur dioxide in aqueous solution by means of iodine was not to be recommended on account of irregularities which are not encountered when the reaction is permitted to take place by allowing the solution of sulphur dioxide to flow into excess of iodine solution, the excess of iodine being determined by sodium thiosulphate.

CHAPTER VI

THE DETERMINATION OF HALIDES BY STANDARD SILVER NITRATE

General

A solution of silver nitrate finds frequent application in volumetric analysis for the determination of soluble halides by precipitation of insoluble silver halide. Soluble chlorides, bromides, and iodides are determined quantitatively by precipitation as the insoluble silver halide. Thus when a solution of silver nitrate is added to a solution of potassium chloride, double decomposition takes place in accordance with the equation

$$AgNO_3 + KCl = AgCl + KNO_3.$$

The end-point may be determined with great accuracy by taking advantage of the property which the silver halides possess of coagulating when agitated; it is thus possible to ascertain when precipitation is complete by adding the reagent very cautiously and observing whether any further precipitation takes place.

A more rapid method of determining the end-point is to add a very small quantity of a solution of potassium chromate to the liquid undergoing titration. As long as precipitation of the halide is incomplete, the liquid remains of a pale yellow colour, but as soon as the slightest excess of the silver solution has been added, the dark red colour of silver chromate imparts a characteristic pink tint to the liquid. In order to employ potassium chromate successfully as an indicator, it is essential that the solution which is undergoing titration should be exactly neutral. In presence of free acid the precipitation of silver chromate is interfered with, while free alkali causes the separation of silver oxide. For most purposes it is advisable to determine the end-point by the aid of potassium chromate; but when great accuracy is aimed at, the slower method of ascertaining when precipitation is just complete will be found more satisfactory. When potassium chromate is used, the end-point is more readily observed by gaslight than it is by daylight.

Preparation of a deci-normal solution

Silver nitrate may be obtained of a high degree of purity by recrystallization. The molecular weight of this salt is 169·89, International atomic weights being used; consequently a deci-normal solution will contain 16·989 grammes or, with sufficient approximation, 17 grammes of the salt dissolved in one litre. A solution of silver nitrate prepared by weighing out the necessary quantity of the salt and diluting it to one litre may be relied upon as being of sufficient exactness for all ordinary purposes, but if the experimenter be in any doubt regarding the purity of the silver nitrate used the solution should be standardized by means of pure sodium chloride.

Instead of making up a standard solution of silver nitrate from the recrystallized salt, the solution may be prepared by starting with pure metallic silver. The weighed quantity of the metal is dissolved in pure dilute nitric acid with the aid of a gentle heat, care being taken to avoid loss of liquid by effervescence. When solution is complete the excess of acid is removed by evaporating the solution to dryness, and the residue is then dissolved in water and made up to the necessary volume.

Determination of halides

As has been explained already, all soluble chlorides, bromides, and iodides may be determined by precipitation by means of standard silver nitrate. If the solution to be titrated contains free acid, and if it be desired to employ potassium chromate as an indicator, it is necessary to neutralize the free acid before titration. In many cases the simplest method of securing neutrality is to add a slight excess of calcium carbonate to the liquid. For example hydrochloric acid may be determined by addition of excess of calcium carbonate, when the hydrochloric acid is converted into an equivalent amount of calcium chloride which possesses a neutral reaction, and which is precipitated quantitatively by silver nitrate

$$2HCl + CaCO_3 = CaCl_2 + H_2O + CO_2,$$
$$CaCl_2 + 2AgNO_3 = Ca(NO_3)_2 + 2AgCl.$$

If n_1 c.c. of the hydrochloric acid solution require n_2 c.c. of silver nitrate for complete precipitation, the weight x of hydrochloric acid per cubic centimetre is obtained by the solution of the equation

$$\frac{n_1 x}{n_2 w} = \frac{\text{equivalent of hydrochloric acid}}{\text{equivalent of silver nitrate}} = \frac{36\cdot5}{170},$$

where w is the weight of silver nitrate dissolved in each cubic centimetre of the standard solution.

Indirect determinations by standard silver nitrate

Not only may chlorides be determined by precipitation with silver nitrate solution, but all substances which can be converted quantitatively into neutral chlorides may be determined in this way. For example the alkali salts of many organic acids are converted on ignition into carbonates; these carbonates may be decomposed by excess of dilute hydrochloric acid, and the excess of acid removed by evaporating the resulting solution to dryness. After extracting with water, the solution of sodium or potassium chloride is determined by titration with standard silver nitrate in the ordinary manner, using potassium chromate as indicator. Instead of converting the carbonates into chlorides however, it is more usual to determine the carbonates by titration with standard mineral acid, methyl orange being employed as an indicator.

Determination of two halides in a mixture

A very important application of silver nitrate as a volumetric reagent consists in the determination of two halides such as sodium chloride and potassium chloride or potassium chloride and potassium bromide when present together in solution and the total weight of the mixed salts in a given volume of solution is known. The method depends upon the difference between the equivalent weights of the two halides, and the greater this difference, the more accurate will be the determination. A single determination with standard silver nitrate is sufficient for the estimation of the amounts of the two constituents of the mixture, as the following example will show.

If we have a mixture of sodium chloride and potassium chloride, the solution containing w_1 grammes of the mixed salts dissolved in one litre, and we find from the results of the titrations that w_2 grammes of silver nitrate are required for complete precipitation per litre, the weights of the two chlorides are obtained by solving the simultaneous equations

$$x + y = w_1 \quad \ldots\ldots\ldots\ldots\ldots\ldots (1),$$

$$\frac{170x}{58\cdot5} + \frac{170y}{74\cdot6} = w_2 \quad \ldots\ldots\ldots\ldots\ldots (2),$$

where x and y denote respectively the weights of sodium chloride and potassium chloride in one litre of the solution.

Again if it is required to determine the weights of potassium chloride and potassium bromide in a solution containing w_3 grammes of the mixed salts dissolved in one litre, aliquot portions of the solution are titrated by means of standard silver nitrate in the usual manner, and if it be found that w_4 grammes of silver nitrate are required for complete precipitation per litre, x the weight of potassium chloride and y the weight of potassium bromide are calculated from the equations

$$x + y = w_3 \quad \ldots\ldots\ldots\ldots\ldots\ldots (3),$$

$$\frac{170x}{74\cdot6} + \frac{170y}{119\cdot1} = w_4 \quad \ldots\ldots\ldots\ldots\ldots (4).$$

It is clear that since the method depends upon the difference between the equivalent weights of the two constituents of the mixture, the accuracy of the process is somewhat limited. The method is most satisfactory when there is not a great difference in the relative amounts of the two constituents of the mixture; it is easy to understand that if a solution containing a relatively large amount of one constituent as compared with the other be titrated with silver nitrate, a very small error in the titration will give rise to a very considerable error in the calculated values.

Theory of errors involved in indirect analysis

Let us consider, as in the special example already discussed, a mixture of two constituents: w_1 grammes of the mixture containing x grammes of a constituent of equivalent weight m_1 and y grammes of a constituent of equivalent weight m_2. Let n be the equivalent weight of the substance employed in solution for the determination, and let w_2 grammes of this substance be required for reaction with w_1 grammes of the mixture; consequently we may write the equations

$$x + y = w_1 \quad\dots\dots\dots\dots\dots\dots(1),$$

and
$$\frac{nx}{m_1} + \frac{ny}{m_2} = w_2 \quad\dots\dots\dots\dots\dots\dots(2),$$

$$y = w_1 - x;$$

$$\therefore \ \frac{nx}{m_1} + \frac{nw_1}{m_2} - \frac{nx}{m_2} = w_2,$$

$$\therefore \ x\left(\frac{n}{m_1} - \frac{n}{m_2}\right) = w_2 - \frac{nw_1}{m_2},$$

or
$$x = \frac{w_2 - \dfrac{nw_1}{m_2}}{\dfrac{n}{m_1} - \dfrac{n}{m_2}}.$$

Differentiating with respect to w_2,

$$\frac{dx}{dw_2} = \frac{1}{n\left(\dfrac{1}{m_1} - \dfrac{1}{m_2}\right)}.$$

This equation shows that the error in x due to unit error in w_2 is inversely proportional to the difference in the reciprocals of m_1 and m_2.

Determination of chlorates

All chlorates are decomposed on ignition into chlorides with evolution of oxygen. For example potassium chlorate decomposes according to the equation

$$2KClO_3 = 2KCl + 3O_2.$$

As a matter of fact the decomposition is not quite so simple as that represented by this equation, since potassium chlorate on heating is converted partly into potassium perchlorate

$$4KClO_3 = 3KClO_4 + KCl.$$

But since the potassium perchlorate on heating is decomposed completely into potassium chloride with evolution of oxygen

$$KClO_4 = KCl + 2O_2,$$

the decomposition of the chlorate may be represented as taking place in accordance with the first equation.

The weighed quantity of the chlorate is ignited in a crucible until constant in weight; the residue is then extracted with water and diluted to some definite volume. Aliquot portions of the solution are then determined by titration with standard silver nitrate.

Determination of alkali cyanides in aqueous solution

If a solution of silver nitrate be added to a solution of potassium cyanide, double decomposition takes place with formation of silver cyanide. But in presence of the excess of the alkali cyanide, the silver cyanide is not precipitated since it is kept in solution as the soluble potassium silver cyanide $KAg(CN)_2$. The reaction may therefore be represented by the equation

$$2KCN + AgNO_3 = KAg(CN)_2 + KNO_3.$$

If the addition of the silver nitrate be continued, a point is at length reached at which silver cyanide is precipitated, that is to say, the following reaction begins:

$$KAg(CN)_2 + AgNO_3 = 2AgCN + KNO_3.$$

The end-point of the reaction shown in the first equation is determined without the addition of any foreign substance to the solution to serve as an indicator, it is simply that point at which the previously clear liquid becomes slightly turbid as the result of the separation of insoluble silver cyanide. It is clear from what has been stated that the calculation of the amount of potassium cyanide in solution is to be made from the first equation given above; or in other words if n_1 c.c. of the potassium cyanide solution require the addition of n_2 c.c. of silver nitrate solution containing w grammes of the salt dissolved in each cubic centimetre, the weight x of potassium cyanide per cubic centimetre is calculated from the equation

$$\frac{n_1 x}{n_2 w} = \frac{2 \times 65}{170}.$$

Owing to the highly poisonous character of hydrocyanic acid, the vapour of which is continuously being given off from solutions of potassium cyanide in consequence of hydrolysis, it is not advisable to measure out potassium cyanide solution in an ordinary pipette; it is much safer to make use of an automatic delivery pipette which obviates the necessity of inhaling the poisonous vapour.

Determination of silver by means of standard sodium chloride

The examples which we have given of the application of silver nitrate to the determination of chlorides, bromides, and iodides in solution may clearly be applied in the reciprocal way. That is to say, silver in its soluble salts may be quantitatively precipitated by means of solutions of any halide. For this purpose use is almost invariably made of sodium chloride, and indeed the determination of silver by means of standard sodium chloride solution is one of the oldest volumetric processes, and is still sometimes employed in the wet-assay of silver. For this latter purpose it is usual to determine the end-point by ascertaining when precipitation is complete, the procedure being rendered very accurate by completing the precipitation with a solution of sodium chloride of one-tenth of the strength of that used for precipitating the main quantity of the metal.

CHAPTER VII

THE DETERMINATION OF SILVER IN ACID SOLUTION BY STANDARD AMMONIUM THIOCYANATE

If a solution of ammonium or potassium thiocyanate be added to a solution of silver nitrate, double decomposition takes place with precipitation of the sparingly soluble silver thiocyanate AgCNS, thus

$$AgNO_3 + NH_4CNS = AgCNS + NH_4NO_3.$$

Upon this reaction, Volhard founded an elegant and accurate method for the volumetric determination of silver. Soluble thiocyanates react with ferric salts with formation of the well-known blood red colour of ferric thiocyanate; and consequently the end-point of the reaction is determined with great precision by adding a small quantity of ferric sulphate to the silver solution. As long as the precipitation of the silver thiocyanate is incomplete the solution remains colourless, but as soon as the slightest excess of thiocyanate has been added, the red colour of ferric thiocyanate imparts a permanent red colour to the liquid, which persists when it is shaken.

Preparation of a deci-normal solution of ammonium thiocyanate

Ammonium thiocyanate being a very deliquescent salt, it is very difficult to weigh out the theoretical quantity (7·6 grammes per litre) to make a solution of exactly deci-normal strength. The best plan is to weigh out a little more than the necessary quantity, so as to make up a solution of rather greater than deci-normal strength. The exact strength of this solution is then found by titration with standard silver nitrate solution. The thiocyanate solution may then be diluted so as to be exactly deci-normal. In carrying out the titrations it is advisable to add a small quantity of nitric acid (free from lower oxides of nitrogen) to the silver solution, in order to decompose any basic ferric salt which would otherwise impart a brown colour to the liquid and thereby render the end-point less sharp than it should be.

If n_1 c.c. of standard silver nitrate containing w grammes of the salt per c.c. require n_2 c.c. of ammonium thiocyanate to complete the reaction, and if x be the weight of ammonium thiocyanate per c.c., we obtain x by solving the equation

$$\frac{n_2 x}{n_1 w} = \frac{\text{equivalent of ammonium thiocyanate}}{\text{equivalent of silver nitrate}} = \frac{76}{170}.$$

Applications

In the previous chapter we have explained the use of standard silver nitrate as a volumetric reagent for the determination of halides. It was also explained that the estimation of silver may be readily effected volumetrically by the use of a standard solution of sodium or potassium chloride. If the end-point is to be found by means of potassium chromate, the liquid undergoing titration and the volumetric precipitant must both be neutral. Ammonium thiocyanate therefore possesses an advantage over sodium chloride as a reagent for the determination of silver, as it is available in the presence of a considerable excess of free acid. Further, it may be employed for the determination of silver in presence of various other metals, including copper up to 70 per cent.; it is therefore of great value in the analysis of silver alloys. As an example we shall describe the procedure for the determination of silver in a silver-copper alloy. A known weight of the alloy (about 0·4 gramme) is dissolved in dilute nitric acid with the aid of a gentle heat, care being taken to avoid loss of liquid by effervescence. After the action has ceased, the contents of the vessel and washings are transferred to a measuring flask and distilled water is added to dilute the contents to the necessary volume. Aliquot portions of the liquid are withdrawn and titrated with standard ammonium thiocyanate solution, a small quantity of a solution of ferric sulphate being added as an indicator.

The quantity of ferric sulphate which is added to a silver

solution which is titrated by ammonium thiocyanate is a matter of some little importance. Too small a quantity should be avoided, as the end-point is sharper in presence of a relatively large quantity of the iron salt. A little practice will soon enable the experimenter to ascertain how much of the indicator to add in order to obtain the best result.

Determination of halides

The determination of chlorides, bromides, or iodides may be effected satisfactorily by means of a standard solution of ammonium thiocyanate by first adding to the solution of the halide a known quantity of silver nitrate solution, which must be in excess, and then determining by residual titration with standard ammonium thiocyanate the amount of silver nitrate remaining unprecipitated. Formerly it was customary to titrate with ammonium thiocyanate in presence of the precipitated silver halide, but it has gradually become recognized that more accurate results are to be obtained by first filtering off the precipitated silver halide, and titrating the filtrate and washings by means of ammonium thiocyanate. The reason of the lower degree of accuracy which is attained by precipitating the silver thiocyanate in presence of the silver halide has been ascribed to the solvent action of ammonium thiocyanate on silver halides.

All of the substances, including the various mixtures the determination of which we have described in the previous chapter, may be determined by the addition of excess of silver nitrate and back-titration with ammonium thiocyanate. In addition, however, the fact that ammonium thiocyanate is available in the presence of free acid renders the method of residual titration by this substance of great availability. The estimation of chlorides which give an acid reaction in consequence of hydrolysis may be effected with satisfactory results.

CHAPTER VIII

ACIDIMETRY AND ALKALIMETRY

Introduction

When aqueous solutions of an acid and a base are allowed to interact, neutralization takes place with formation of a salt and water, according to the general equation

$$\text{Acid} + \text{Base} = \text{Salt} + \text{Water}.$$

The process of neutralization is not always quantitative, that is to say, there are many cases in which the reaction is reversible to some extent. The reverse reaction to neutralization, that is, the partial decomposition of a salt into free acid and free base, is termed hydrolysis. Hydrolysis, theoretically at least, occurs whenever any salt is dissolved in water, but in those cases in which the salt is the product of a strong acid, such as hydrochloric, and a strong base, such as sodium hydroxide, the degree of hydrolysis is practically infinitesimal. It is otherwise when the salt is the product of an acid and a base, one or both of which is weak. If the acid is weak and the base is strong, the result of hydrolysis is that the salt in solution has an alkaline reaction. Sodium acetate, for example, reacts alkaline in aqueous solution. On the other hand, if the salt is the product of a weak base and a strong acid, the salt will possess an acid reaction in solution in consequence of hydrolysis. Ferric chloride, for example, reacts strongly acid when dissolved in water. The cases in which the salt is derived from a weak base and a weak acid are somewhat more complicated, but for the present purpose it is unnecessary to consider them, since it is impossible to determine a weak acid by means of a weak base satisfactorily, and, as we shall see later, it is never necessary to do so; and generally, in acidimetry and alkalimetry methods are so chosen that hydrolysis comes as little into play as possible.

Indicators

In order to determine the point of neutrality, a small quantity of a suitable "indicator" is added to the liquid undergoing titration. The indicators in common use are organic compounds which possess different colours in acid and in alkaline solution. The indicators themselves are feebly acidic or basic substances, and the slightest excess of either acid or base is sufficient to determine the change of colour. The choice of a suitable indicator to be employed in any particular acidimetric or alkalimetric determination is a matter of great importance, and is to be made from a consideration of the properties of the indicator and the relative strengths of the acid and base which are to interact. The theory of indicators will be discussed more fully in the next chapter; for the present purpose it will be sufficient to describe the properties of some of the various indicators in common use and to indicate the reasons for making any particular choice.

Litmus

This substance is one of the oldest indicators, and for many purposes it gives very satisfactory results. It is a substance of a feebly acid nature, and is blue in alkaline and red in acid solution. Neutrality is indicated by a pale lavender tint. This indicator may be employed in the titration of strong acids by strong bases, but it is not a sufficiently feeble acid to give satisfactory results with the weakest organic acids. Litmus is a weaker acid than carbonic acid; in other words, it undergoes a colour change under the influence of carbon dioxide; consequently in titrating alkaline carbonates in presence of litmus, it is necessary to keep the solution of the carbonate at the boiling point in order to expel carbon dioxide from the solution.

Phenolphthalein

Phenolphthalein or dihydroxyphthalophenone is a substance having the formula

$$C_6H_4 \quad C_6H_4OH$$
$$CO \quad C\!-\!C_6H_4OH$$
$$O$$

and is so weak an acid as to be almost devoid of acid properties in solution. It is soluble in dilute alcohol, and is red in alkaline but colourless in acid solution. This indicator, on account of its excessively weak acidic properties, is very well adapted for the determination of the weakest organic acids by titration with caustic soda. Phenolphthalein cannot be used for the titration of weak bases like ammonia, for reasons which will be explained in the next chapter, nor may it be used in presence of ammonium salts. This indicator is sensitive to carbon dioxide; if an alkaline carbonate be titrated by a strong acid in presence of phenolphthalein, neutrality will be indicated when the normal carbonate is converted into the bicarbonate, that is, when the normal carbonate is half neutralized.

Methyl orange

This substance is the sodium salt of dimethylaminoazobenzene-sulphonic acid $SO_3HC_6H_4N_2C_6H_4N(CH_3)_2$. It is soluble in water, and is yellow in alkaline but red in acid solution. The substance is possessed of fairly well defined acid properties, and is an extremely good indicator for strong acids. Since methyl orange is a stronger acid than such acids as carbonic and boric, alkaline carbonates and borates may be titrated by means of strong acids with very good results, as this indicator is practically unaffected by weak acids. Methyl orange also yields satisfactory results when weak bases like ammonia are titrated by means of strong acids. In using this indicator it is important to avoid the use of too much of it, one drop of a solution of one-tenth per cent. strength is usually amply sufficient. If too great a quantity be employed the end-point will not be sharply defined. The solution to be titrated should not be too dilute, or else intermediate shades of colour will be obtained as the liquid approaches neutrality. Some persons appear to have considerable difficulty in obtaining satisfactory results with methyl orange, but it must be borne in mind that some eyes are much more sensitive to the colour change than others.

Methyl red

This substance is closely related to methyl orange, but contains an ortho carboxylic instead of a para sulphonic acid group. The substance has the formula $COOHC_6H_4N_2C_6H_4N(CH_3)_2$. It is possessed of somewhat weaker acid properties than methyl orange, and is yellow in alkaline but violet-red in acid solution. It is an extremely sensitive indicator and is very well adapted for the titration of weak bases by strong acids. The colour change is very much more striking than that of methyl orange, and is preferable to the latter indicator on that ground. On the other hand it is more sensitive to carbon dioxide.

Paranitrophenol

This substance has the formula $C_6H_4(OH)(NO_2)$. It is sparingly soluble in water, but more soluble in water containing alcohol. In acid solution it is perfectly colourless, but in alkaline solution it is coloured yellow. In general properties this indicator has much in common with methyl orange, but is rather more sensitive to carbon dioxide. Like methyl orange, it gives satisfactory results when weak bases are titrated by means of strong acids.

Other indicators

Many other organic compounds which are possessed of different colours in acid and in alkaline solution have been proposed from time to time as indicators for special purposes. It will be sufficient to mention such indicators as lacmoid, phenacetolin, congo red, and rosolic acid, but we must refer the reader to some larger treatise for an account of the properties of these substances.

Standard solutions of acid and alkali

In all acidimetric and alkalimetric work it is obviously necessary to have some one substance of great purity with which to prepare a solution of accurately known strength, and in terms of which all other acid and alkaline solutions may be readily standardized. Various substances have been suggested for this purpose.

Succinic acid $C_2H_4(COOH)_2$ is a substance which may be obtained of a high degree of purity by recrystallization; a solution of this substance may be employed to determine the exact strength of solutions of potassium or sodium hydroxide. Since succinic acid is a weak acid, phenolphthalein must be employed in titrating with this acid. Again sodium carbonate Na_2CO_3 may be obtained in a very pure state by heating the bicarbonate $NaHCO_3$ at a temperature of about 270° C., when water and carbon dioxide are expelled. A standard solution of sodium carbonate may readily be prepared and will serve as a reliable standard for the determination of the strengths of acids such as sulphuric or hydrochloric. Some operators employ borax purified by recrystallization as standard substance. The standardization of acids may also be effected with very satisfactory results by the use of metallic sodium. This metal can readily be obtained in a state of great purity, and for most purposes there is no better method of standardizing acids for volumetric work. The metal is cut into small pieces, each of the order of one gramme in weight, freed from naphtha by pressing between filter paper, and after weighing rapidly to the nearest milligramme, each piece is dropped into a flask containing some rectified alcohol. The metal dissolves with evolution of hydrogen and formation of sodium ethoxide which remains dissolved in the excess of alcohol. The flasks should be held in an inclined position during the solution of the metal to prevent loss of liquid by effervescence. The reaction which takes place is represented by the equation

$$2Na + 2C_2H_5OH = 2C_2H_5ONa + H_2.$$

When the metal has been completely dissolved, excess of water is added, when sodium hydroxide is formed according to the equation

$$C_2H_5ONa + H_2O = C_2H_5OH + NaOH.$$

The acid which is to be standardized is then titrated against this solution, litmus being used as an indicator. In standardizing an acid by this method, it is important to weigh the metal rapidly to avoid atmospheric oxidation as much as possible, and it is advisable to make successive determinations by weighing out separate quantities of sodium, rather than by weighing out one piece of the metal and making up the resulting solution of sodium hydroxide to some definite volume by dilution with water. The sodium must be dissolved in rectified alcohol; methylated spirit must not be used, as the impurities in this substance frequently interfere with the accuracy of the experiment. It is perhaps scarcely necessary to remark that the sodium must not be placed in water directly. In calculating the strength of the acid which is being standardized, it is to be remembered that 23 parts by weight of sodium are equivalent to 36·5 parts by weight of hydrochloric acid, or to 63 parts by weight of nitric acid or to 49 parts by weight of sulphuric acid. There is no necessity to calculate as an intermediate step the weight of sodium hydroxide which is produced from the weighed quantity of metal.

An ingenious method of standardizing acids for volumetric work depending upon the use of Iceland spar was devised some years ago by Orme Masson (*Chem. News*, 1900, p. 73). Iceland spar is calcium carbonate $CaCO_3$ probably in as pure a state as it is possible to obtain any substance. This substance dissolves in hydrochloric acid with formation of calcium chloride and evolution of carbon dioxide according to the equation

$$CaCO_3 + 2HCl = CaCl_2 + H_2O + CO_2.$$

The essential feature of Masson's method of employing this substance for standardizing hydrochloric acid is the use of a known weight of the substance which is always in excess of the amount theoretically required to react with the quantity of hydrochloric acid taken in any particular determination and determining gravimetrically the amount of Iceland spar remaining unacted upon. Small compact fragments of Iceland spar are broken from a large rhomb of the substance and a suitable weight (from 2 to 3 grammes) of the fragments placed in a beaker or other suitable vessel. The beaker with its contents is then carefully weighed. It is to be noted that it is unnecessary to know the weight of either beaker or Iceland spar separately. The measured quantity of hydrochloric acid is then run in from a burette, precautions being taken to guard against loss of liquid by effervescence. The beaker with its contents is then set aside until all action has ceased. Then after boiling for some time to expel dissolved carbon dioxide from the solution, distilled water being added from time to time to prevent the solution from becoming too much concentrated, the perfectly clear and neutral solution of calcium chloride is decanted off from the excess of Iceland spar which remains in the form of compact fragments. The beaker with the residual solid is then washed out several times with distilled water, dried at a temperature of about 110° C. and weighed. The difference between the initial and final weights of the beaker and Iceland spar is clearly equal to the weight of spar which has been dissolved by the volume of hydrochloric acid taken in the experiment.

Since the molecular weight of calcium carbonate is very nearly 100, it follows that 20 c.c. of a strictly normal acid should dissolve exactly one gramme of Iceland spar. The calculation of the strength of the acid is therefore effected with a minimum of arithmetical work, and further the method has the advantage of extreme simplicity of working, only one volume measurement and two weighings being involved in each determination.

It might be supposed that the method was liable to yield inaccurate results, since it is conceivable that a portion of the residual spar might dissolve in the form of soluble calcium bicarbonate $Ca(HCO_3)_2$. If such were the case, however, the bicarbonate would be decomposed by boiling the solution with formation of the normal carbonate and evolution of carbon dioxide. However, the author states that no turbidity was observable on boiling the solution, and consequently the solid must dissolve entirely as calcium chloride. The experiments which were described certainly indicate that the method is capable of yielding results of extreme accuracy.

An accurate method of preparing standard hydrochloric acid by passing pure dry gaseous hydrogen chloride into water and weighing the amount absorbed was devised by Moody (*Trans. Chem. Soc.*, 1898, p. 658). The hydrogen chloride is generated by heating rock salt with concentrated sulphuric acid, or by dropping strong sulphuric acid into ordinary concentrated hydrochloric acid. In the latter case it is necessary to dry the gas by strong sulphuric acid. The method is a simple one to carry out, and the results obtained show that it leaves nothing to be desired from the point of view of accuracy.

Having standardized a solution of hydrochloric or sulphuric acid by means of sodium or by other suitable means, we are in a position to prepare standard solutions of all other acids and alkalis*. For example, suppose we have determined the exact strength of a solution of hydrochloric acid and we wish to prepare a quantity of nitric acid of normal strength, all that is necessary is to refer to tables showing the relation between the specific gravity of aqueous solutions of nitric acid and the strength of the acid. Pure concentrated nitric acid is then diluted with water, until the mixture when cold has approximately the correct specific gravity, or preferably a specific gravity very little above that possessed by the normal acid. A solution of caustic soda, the strength of which need not be known, is taken and titrated first by the standard hydrochloric acid, and then by the nitric acid. Since one gramme molecular weight of sodium hydroxide is equivalent to 36·5 grammes of hydrochloric acid and also to 63 grammes of nitric acid, it is a simple matter to calculate the exact strength of the nitric acid. The nitric acid is then cautiously diluted with water and restandardized, and the process repeated until the strength of the acid is exactly normal.

In standardizing the weak organic acids such as acetic and tartaric, it is essential to employ phenolphthalein as the indicator. On the other hand if it is required to prepare a standard solution of ammonia, methyl orange or methyl red must be used in titrating this substance.

The specific gravity of solutions of substances such as sulphuric acid and sodium hydroxide varies in a perfectly regular manner with the concentration of the solution. Consequently the specific gravity is a valuable aid during the preliminary process of preparing a standard solution. In certain cases, however, notably in the case of hydrochloric acid, standard solutions of approximately known strength may be prepared by taking advantage of the

* It will be clear that in acidimetry and alkalimetry it is only necessary to have *one* standard liquid of each kind. Hydrochloric acid and sodium hydroxide are the reagents usually employed; the former has the advantage over other acids that its strength can be checked independently with reference to silver nitrate.

peculiar phenomena which take place when the aqueous solutions are distilled. It was shown by Roscoe and Dittmar that when a concentrated solution of hydrochloric acid was distilled, the solution gradually became weaker until the strength became constant. On the other hand a dilute solution became more concentrated on distillation, finally becoming of constant strength. The particular strength of hydrochloric acid which behaved on distillation like a pure liquid was found to depend upon the pressure. At a pressure of 760 mm. the constant boiling mixture of hydrochloric acid and water was found to contain 20·2 per cent. of acid. An acid of this strength may be diluted to any desired extent, but the exact strength should in all cases be determined by titration.

Determination of the water of crystallization in hydrated sodium carbonate

It has been explained already that alkaline carbonates may be accurately titrated by means of strong acids, methyl orange being used as an indicator. A few grammes of the crystals are weighed out, dissolved in water, and made up to a suitable volume with water. Aliquot portions of the solution are then titrated by means of standard acid. The weight of anhydrous sodium carbonate is then calculated from the equation

$$Na_2CO_3 + 2HCl = 2NaCl + H_2O + CO_2.$$

It is clear from this equation that 53 grammes of anhydrous sodium carbonate are equivalent to 36·5 grammes of hydrochloric acid.

If w_1 grammes of the crystals are taken, and it is found by titration that w_2 grammes of anhydrous sodium carbonate are present, the weight of combined water is clearly $(w_1 - w_2)$ grammes. This quantity of water has entered into combination with w_2 grammes of anhydrous sodium carbonate; therefore the weight of water combined with one molecular weight of sodium carbonate is $106 \dfrac{(w_1 - w_2)}{w_2}$ grammes. Dividing this fraction by the molecular weight of water (18), we obtain the number of molecules of water of crystallization in the substance.

*Determination of sodium carbonate and hydroxide when
present together*

This estimation is a problem of daily occurrence in alkali
works. Formerly it was the usual practice to determine the
caustic alkali alone by titration with standard acid after separation
of the carbonate by precipitation as insoluble barium carbonate, and
then to determine the total alkali by a separate titration. At the
present time, the determination of the two constituents is effected
by the use of two indicators. If to a solution of a mixture of
alkaline carbonate and hydroxide hydrochloric acid be added in
presence of phenolphthalein, neutrality will be indicated when the
sodium hydroxide has been completely neutralized and the sodium
carbonate converted into sodium hydrogen carbonate. In other
words the neutralization of the two constituents of the mixture
may be represented by the equations

$$NaOH + HCl = NaCl + H_2O,$$
and $$Na_2CO_3 + HCl = NaCl + NaHCO_3.$$

During the titration with phenolphthalein, the tip of the burette
should be kept immersed in the liquid to prevent the escape of
carbon dioxide. A drop of methyl orange is now added to the
liquid, and the titration continued until the reaction is completed.
This second reaction consists in the decomposition of the sodium
bicarbonate into sodium chloride, carbon dioxide, and water

$$NaHCO_3 + HCl = NaCl + H_2O + CO_2.$$

If n_1 c.c. of acid are required to discharge the red colour of the
phenolphthalein, and a further n_2 c.c. of acid are required to change
the yellow colour of the methyl orange to pink, it is clear that the
total alkali is represented by $(n_1 + n_2)$ c.c. of acid. The sodium
carbonate alone corresponds to $2n_2$ c.c. of acid, while the sodium
hydroxide is represented by $(n_1 - n_2)$ c.c. of acid.

Determination of sodium carbonate and bicarbonate in a mixture

From what has been stated in the preceding section, it is clear
that it should be possible to determine sodium bicarbonate and
normal carbonate by the use of two indicators; and, as a matter

of fact, satisfactory determinations of relatively small quantities of bicarbonate in presence of larger quantities of the normal carbonate may be readily carried out by this method. Phenol-phthalein is added to the measured quantity of the solution, and standard hydrochloric or other strong acid is run in until the colour disappears. If v_1 c.c. of acid are added, it is clear that the sodium carbonate is represented by $2v_1$ c.c. of acid. If now a drop of methyl orange be added to the same liquid, and the addition of acid continued until the yellow liquid becomes red, v_2 c.c. of acid run in from the completion of the phenolphthalein titration, that is the total titration is $(v_1 + v_2)$ c.c. of acid, which represents the total alkali. The amount of acid which corresponds to the sodium bicarbonate is clearly equal to the difference between that required for the total alkali and that required for the normal carbonate, or $(v_1 + v_2 - 2v_1)$ c.c. or $(v_2 - v_1)$ c.c.

Determination of ammonium salts

All ammonium salts are decomposed when boiled with excess of solutions of caustic alkalis. Thus when ammonium sulphate is boiled with excess of sodium hydroxide, the following reaction takes place:

$$(NH_4)_2SO_4 + 2NaOH = Na_2SO_4 + 2H_2O + 2NH_3.$$

The ammonia being volatile is expelled from the solution. The determination of an ammonium salt may be effected by adding a *known* excess of a *standard* solution of sodium hydroxide, boiling until all the ammonia has been driven off, and then determining by titration with standard sulphuric or other strong acid the amount of alkali remaining in excess. The amount of ammonium salt present is represented by the difference between the amount of sodium hydroxide originally taken and that found by titration after boiling off all the ammonia. Another method of carrying out the determination is to add an arbitrary excess of caustic soda of unknown strength to the ammonium salt and to conduct the decomposition of the ammonium salt in an apparatus for collecting the evolved ammonia in a measured excess of standard acid (Fig. 4). The amount of acid remaining unneutralized is then determined by titration with standard caustic soda. In this case the amount of ammonium salt present is represented by the difference between the amount of standard acid taken originally and the amount remaining after absorption of the whole of the ammonia resulting from the decomposition of the salt.

It must be borne in mind that all titrations in presence of ammonium salts must be carried out with litmus or preferably with methyl orange or with methyl red. Phenolphthalein must not be used on account of the hydrolysis of the ammonium phenolphthalein salt.

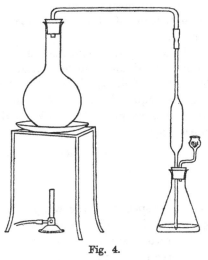

Fig. 4.

Both methods of determining ammonium salts give equally satisfactory results in the absence of substances which introduce complications; but in certain cases, the method of estimating the ammonium salt by collecting the evolved ammonia in excess of standard acid must be employed. For example if it be desired to determine the percentage of the ammonium group in ammonium ferrous sulphate, the addition of caustic soda to this salt is immediately followed by the precipitation of ferrous hydroxide which would make titration with acid impossible. On the other hand the determination may be carried out with satisfactory results by the second method.

Determination of two acids in a mixture when the total acid is known

The general principle of the method of indirect analysis depending upon the difference between the equivalent weights of the two constituents has been already described in connexion with the determination of two halides by means of standard silver nitrate (Chapter VI). It is obvious that a precisely similar procedure should suffice for the determination of two acids such as nitric and sulphuric by means of standard sodium hydroxide, and as a matter of fact satisfactory determinations of this kind may be readily made.

If we have a mixture of nitric acid and sulphuric acid of such a strength that one litre of the solution contains w_1 grammes of the mixed acids, and we find by titration that w_2 grammes of sodium hydroxide are required for complete neutralization of a litre of the solution, the weights of the two acids are found by solving the simultaneous equations

$$x + y = w_1 \ \ldots\ldots\ldots\ldots\ldots\ldots (1),$$

$$\frac{40x}{63} + \frac{40y}{49} = w_2 \ \ldots\ldots\ldots\ldots\ldots\ldots (2),$$

where x and y denote respectively the weights of nitric and sulphuric acid in grammes dissolved in one litre of the mixture.

The theory of the errors involved in indirect determinations of this kind has been already discussed (Chapter VI), and it is clear that the method is open to the objections there indicated.

In the example which we have discussed any indicator may be employed for determining neutrality since both of the constituents of the mixture are strong acids, and caustic soda is a strong base. On the other hand if either or both of the acids to be determined are weak, phenolphthalein must be used as the indicator.

Determination of orthophosphoric acid

This substance belongs to the group of acids of medium strength. The formula of the acid, H_3PO_4, and the fact that it gives rise to three series of salts, characterize it as a tribasic acid. On the other hand, its behaviour on neutralization by caustic soda in presence of various indicators is somewhat peculiar. If the acid be neutralized by sodium hydroxide in presence of methyl orange, neutrality will be indicated when only one of the three hydrogen atoms has been replaced by sodium, dihydrogen sodium phosphate being formed according to the equation

$$H_3PO_4 + NaOH = NaH_2PO_4 + H_2O.$$

On the other hand if phenolphthalein be employed as indicator in the titration of this acid, the addition of sodium hydroxide may be continued until two equivalents of the base have been added to each molecule of phosphoric acid; in other words neutrality will be indicated when hydrogen disodium phosphate has been formed as indicated by the equation

$$H_3PO_4 + 2NaOH = Na_2HPO_4 + 2H_2O.$$

The theoretical explanation of this behaviour on neutralization will be given in the next chapter; for the present purpose all that it is necessary to bear in mind is that the calculation of the amount of acid present from the titration is to be made from the first equation if methyl orange is employed as the indicator, and from the second equation if the determination is carried out in presence of phenolphthalein.

Determination of citric acid

Citric acid is a tribasic acid of the formula

$$CH_2 . COOH$$
$$|$$
$$C(OH)COOH$$
$$|$$
$$CH_2 . COOH$$

It crystallizes with one molecule of water. Like phosphoric acid it behaves differently with different indicators on neutralization with sodium hydroxide. With methyl orange citric acid behaves as a monobasic acid, neutrality being indicated when the formation of the monosodium citrate has taken place. On the other hand, when phenolphthalein is employed as indicator, the trisodium salt of the acid is formed; that is, the substance then behaves as a tribasic acid.

Determination of borax

Borax is a substance of the formula $Na_2B_4O_7$. It is the salt of an extremely weak acid and a strong base. Borax crystallizes with ten molecules of water, and is moderately soluble in water. The aqueous solution reacts strongly alkaline as a result of hydrolysis, and may be titrated by means of a strong acid using methyl orange as an indicator, when free boric acid is formed, but owing to the very feebly acid character of this substance the end-point with methyl orange is quite satisfactory. The neutralization may be expressed by the equation

$$2HCl + Na_2B_4O_7 + 5H_2O = 2NaCl + 4H_3BO_3.$$

If now to this liquid a little phenolphthalein be added, the free orthoboric acid may be titrated by means of sodium hydroxide. The end-point however is quite unsatisfactory, the red colour appearing before the whole of the free boric acid has been neutralized. It has been found, however, that if a moderately large quantity of glycerol or mannitol be added to the liquid undergoing titration, then the end-point does really represent the point at which all the boric acid has been neutralized. The reaction is represented by the equation

$$4H_3BO_3 + 4NaOH = 4NaBO_2 + 8H_2O,$$

sodium metaborate being formed.

It will be observed from what has been stated that borax may be titrated first as a base and then as an acid after liberation of the orthoboric acid by a strong acid, and exactly double the molecular proportion of sodium hydroxide is required for neutralization as acid to the quantity of hydrochloric acid which is required for neutralization as base. The extremely weakly acid properties of boric acid are shown by the fact that when neutralized by caustic soda in presence of phenolphthalein, the end-point is reached while a relatively considerable proportion of the acid remains unneutralized. The manner in which certain polyhydric alcohols such as glycerol stimulate the activity of the boric acid to such an extent that the acid may be quantitatively determined by sodium hydroxide in presence of phenolphthalein is unknown.

The sodium hydroxide which is employed for titrating borax must be absolutely free from carbonate, in the first place because phenolphthalein is being employed as indicator, and secondly because sodium carbonate itself reacts with boric acid in accordance with the equation

$$Na_2CO_3 + 4H_3BO_3 = Na_2B_4O_7 + CO_2 + 6H_2O.$$

In this equation four atoms of boron are equivalent to two atoms of sodium, whereas in the previous equation four atoms of boron are equivalent to four atoms of sodium.

Baryta water as standard alkali

For many purposes it is useful to employ a solution of barium hydroxide as a standard alkaline volumetric reagent. This substance is a strong base and is on that account well adapted for the titration of weak organic acids such as succinic in presence of phenolphthalein. The solution should be of deci-normal strength, since stronger solutions are liable to separation of the solid on account of its limited solubility. Baryta water must be protected very carefully from atmospheric carbon dioxide since this gas causes precipitation of insoluble barium carbonate, which would of course alter the strength of the solution. The solution should therefore be preserved in a bottle in direct connexion with the burette, and a soda lime tube must be inserted in the upper part of the vessel to absorb atmospheric carbon dioxide. Notwithstanding these precautions the solution will require frequent titration with standard acid.

CHAPTER IX

THE THEORY OF INDICATORS

It is difficult, if not impossible, to define in a brief formula what an acid or a base is; that is to say, to give such definitions of these two classes of substances as would be intelligible to anyone who had no practical acquaintance with them. The hypothesis of electrolytic dissociation has greatly facilitated the precise characterization of these classes of compounds. In terms of this hypothesis, acids are defined as compounds of hydrogen which produce free hydrogen ions in aqueous solution; the acidic properties of the solution being associated with the presence of hydrogen ions. Similarly the alkaline properties associated with solutions of basic substances are, in terms of the ionic theory, due to the presence of free hydroxyl ions.

Indicators are to be regarded either as weak acids or as weak bases, and Ostwald has applied the ionization hypothesis in a simple and ingenious manner to the elucidation of the behaviour of the indicators which are commonly applied to determine neutrality. This theory depends upon the following points in the ionic theory:

(1) Strong acids and strong bases are largely dissociated electrolytically in solution; that is, their solutions contain a large quantity of hydrogen or of hydroxyl ions respectively.

(2) Weak acids and weak bases are but little dissociated; their solutions contain only small quantities of hydrogen or of hydroxyl ions respectively.

(3) Salts which are the product of either a weak base and a strong acid, or of a strong base and a weak acid, are largely dissociated in aqueous solution.

(4) Salts belonging to the types considered in (3) undergo hydrolytic dissociation as well as ionization. Hydrolysis is still more pronounced in the cases of salts derived from weak bases and weak acids. Such solutions contain besides small quantities

of the cations and anions of the salt, a large quantity of the salt
in the form of undissociated acid and base. For example phenolphthalein is an extremely weak acid. The addition of ammonium
hydroxide to this substance results in the formation of the
ammonium salt of phenolphthalein, a salt which is hydrolysed
to a very considerable extent. Denoting, for shortness, phenolphthalein by HPh, the formation of ammonium phenolphthalein
may be represented by the equation

$$NH_4OH + HPh \rightleftharpoons NH_4Ph + H_2O.$$

Whether ionization or hydrolysis will predominate in any
particular case depends of course on the conditions of equilibrium.

(5) If ions, which are capable of giving rise to a feebly
dissociated compound, happen to come together in solution,
that compound is formed. For this reason, and also for another
reason which will appear presently, the already small electrolytic
dissociation of a weak acid or a weak base is reduced to a further
extent if excess of hydrogen ions or of hydroxyl ions be added to
the solution. In the case of a weak organic acid, such as acetic
acid, there is a definite equilibrium between the undissociated part
of the molecule and the ions

$$CH_3COOH \rightleftharpoons \overset{+}{H} + CH_3\overset{-}{C}O_2.$$

This equilibrium is regulated by the law of mass action as represented by the equation

$$a \cdot b = k \cdot c,$$

where a denotes the concentration of the hydrogen ions, b the
concentration of the anion, c the concentration of the undissociated
molecule, and k the equilibrium constant.

If, now, a strong acid, in other words if excess of hydrogen ions,
be added to the solution, the equilibrium will be disturbed with
the result that the ionization of the weak acid will be suppressed.

We can now discuss the behaviour of an indicator towards acids
and bases from the standpoint of this theory. Denoting as before
phenolphthalein by the symbol HPh we may write the equilibria

$$\overset{+}{H} + \overset{-}{Ph} \rightleftharpoons HPh \dots\dots\dots\dots\dots (1),$$

$$\overset{-}{OH} + \overset{+}{K} \rightleftharpoons KOH \dots\dots\dots\dots\dots (2).$$

The result of adding potassium hydroxide to the phenolphthalein is to form water by the union of the hydrogen and hydroxyl ions. Now water is dissociated electrolytically only to an exceedingly small extent. The result is that hydrogen ions disappear from the solution. The equilibrium represented by equation (1) is destroyed and consequently, in order to establish equilibrium again, more molecules of phenolphthalein must undergo ionic dissociation. The result of this is that the concentration of the phenolphthalein ions is increased.

The solution is now coloured red, so we conclude that the red colour is to be connected with the phenolphthalein ion. The condition of the solution is now represented by

$$\overset{+}{K} + \overset{-}{Ph} \rightleftharpoons KPh \ \dots\dots\dots\dots (3),$$

$$\overset{-}{OH} + \overset{+}{K} \rightleftharpoons KOH \ \dots\dots\dots\dots (4).$$

It is clear that in consequence of the addition of a sufficient quantity of caustic potash, the hydrogen ions of the phenolphthalein and the hydroxyl ions of the potassium hydroxide combine to form water so that instead of the weak, feebly dissociated acid phenolphthalein, we now have the potassium salt of phenolphthalein which, being a salt derived from a weak acid and a strong base, is highly ionized.

If now we add a strong acid, such as hydrochloric, to the red alkaline solution, the excess of alkali is neutralized according to the equation

$$\overset{-}{OH} + \overset{+}{K} + \overset{+}{H} + \overset{-}{Cl} \rightleftharpoons \overset{+}{K} + \overset{-}{Cl} + H_2O \ \dots\dots (5).$$

Finally, if excess of hydrochloric acid be added, in other words if an excess of hydrogen ions be added to the solution, the solution will contain two kinds of ions which are capable of producing a compound which is very slightly ionized in solution, viz. phenolphthalein. The formation of phenolphthalein may be represented by the equation

$$\overset{+}{K} + \overset{-}{Ph} + \overset{+}{H} + \overset{-}{Cl} \rightleftharpoons \overset{+}{K} + \overset{-}{Cl} + HPh \dots\dots (6),$$

or more simply thus

$$\overset{-}{Ph} + \overset{+}{H} \rightleftharpoons HPh \dots\dots\dots\dots (7).$$

Since phenolphthalein ions disappear from the solution, in order to restore the equilibrium represented by equation (3) the process must take place from right to left, *i.e.* the small quantity of undissociated potassium phenolphthalein dissociates into its ions, and the phenolphthalein ions combine with the excess of hydrogen ions of the hydrochloric acid to form undissociated phenolphthalein. In other words the end of the reaction consists in a complete reformation of the undissociated phenolphthalein molecule, so that we return to the condition of equilibrium represented in equation (1)

$$\overset{+}{H} + \overset{-}{Ph} \rightleftharpoons HPh$$

and consequently to a colourless solution.

What has been stated with regard to strong acids such as hydrochloric is equally true of weak organic acids, since they are sufficiently strong to suppress the ionization of phenolphthalein so that a colourless solution is the result.   Now experiment shows that strong bases such as potassium and sodium hydroxides may be determined with great accuracy by titration with strong acids, but weak bases such as ammonia cannot be determined in this way.   The explanation of this is given by the theory which we have discussed.   After the strong base has been neutralized by the strong acid, the solution contains the ions of potassium and phenolphthalein, which result from the dissociation of the potassium salt of phenolphthalein; and an extremely small concentration of hydrogen ions is required in order to suppress the phenolphthalein ions into undissociated phenolphthalein, that is to decolorize the solution.

But if the base is a weak one like ammonia, the strongly hydrolysed salt ammonium phenolphthalein is formed.   The red colour of the solution is due to a large excess of ammonia which suppresses hydrolysis.   If this excess of ammonia is removed by titration with acid, a point is at length reached when there is not enough ammonia to suppress the hydrolysis of the salt.   The solution does not contain the ions of phenolphthalein but colourless undissociated molecules of phenolphthalein; in other words the red colour is discharged before all the ammonia is neutralized by the acid which is employed in the titration.

Methyl orange

This substance is a moderately strong acid; in dilute solution it is ionized to a considerable extent, the ion is yellow and the undissociated molecule is red

$$\overset{+}{H} + \overset{-}{Me} \rightleftarrows HMe.$$

Strong acids are capable of suppressing this electrolytic dissociation with the result that the solution turns red. But the addition of a weak acid does not bring a sufficient excess of hydrogen ions into the solution to suppress the ionization of the methyl orange, so that no sharp colour change takes place. In other words methyl orange is useless for the titration of weak acids.

The addition of alkali turns methyl orange yellow because the moderately strong acid methyl orange forms salts with weak bases which are dissociated electrolytically to a sufficient extent to give the yellow colour of the ion of methyl orange.

If now the yellow coloured ammonia solution be titrated with a strong acid, like hydrochloric, the solution will remain yellow until only a very small quantity of ammonia is left. There is no difficulty due to hydrolysis with this indicator, because the salt which is formed, ammonium methyl orange, is the product of a weak base and a fairly strong acid. The colour change from yellow to red, which corresponds with the complete suppression of ionization, the formation of the molecule of methyl orange, first occurs when there is a drop of acid in excess.

Dissociation constants of indicators

It has been shown by Ostwald that the ionization of a weak electrolyte is strictly in accordance with the law of mass action, that is to say, the equilibrium between the undissociated molecule of a weak acid and its ions is regulated by the equation

K × (concentration of undissociated molecule)

= (concentration of hydrion) × (concentration of anion).

In the special case in which the indicator is ionized to the extent of 50 per cent., we have the simple relation, $K = (\overset{+}{H})$*, that is to say, the dissociation constant of an indicator is equal to the concentration of the hydrogen ions in the solution in which the indicator is ionized to the extent of 50 per cent.

The dissociation constants of a number of the indicators in common use have been determined. For a description of the experimental methods by which the dissociation constants have been determined we must refer the reader to original papers.

The following results have been obtained by Salm (*Zeitsch. physikal. Chem.*, 1906, **57**, p. 471).

	Indicator	K Dissociation constant
Acids	Methyl orange	$4\cdot6 \times 10^{-4}$
„	Methyl red†	$1\cdot05 \times 10^{-5}$
„	Paranitrophenol	$2\cdot3 \times 10^{-7}$
„	Rosolic acid	$1\cdot1 \times 10^{-8}$
„	Alizarine	$8\cdot8 \times 10^{-9}$
„	Phenolphthalein	$8\cdot0 \times 10^{-10}$‡
Bases	Cyanine	$4\cdot2 \times 10^{-6}$
„	Dimethylaminoazo-benzene	$1\cdot4 \times 10^{-11}$

A knowledge of the dissociation constants of indicators furnishes us not only with the knowledge of which indicator is the best to employ for any particular titration, but also the error which a less suitable indicator would have introduced. It is essential in any given acidimetric or alkalimetric determination to titrate to a particular colour corresponding to a given concentration of hydrogen ions.

If a strong acid is titrated by a strong base, the resulting solution will possess a neutral reaction. It is otherwise if one of the components of the salt is a weak acid or a weak base. If we add to a solution of ammonia the equivalent quantity of hydrochloric acid, the resulting solution will react acid in consequence of hydrolysis. If we add to boric acid the equivalent quantity of sodium hydroxide the resulting solution will be strongly alkaline in reaction. Now an indicator ought to tell us when a given

* Or $K = (\overset{-}{OH})$ in the case of basic indicators.

† Tizard (*Trans. Chem. Soc.*, 1910, **97**, p. 2477).

‡ The values given for the dissociation constant of phenolphthalein by different investigators vary considerably. Values higher than 10^{-9} have been given.

quantity of acid has had the equivalent quantity of base added to it; and this point, which we shall term the equivalent point, is only identical with the neutralization point in the case of strong electrolytes.

Behaviour of polybasic acids

In the previous chapter we had occasion to mention the peculiar behaviour of citric and orthophosphoric acids towards phenolphthalein and methyl orange. What explanation of the difference in behaviour towards the two indicators is forthcoming? To this question a perfectly general answer can be given. A polybasic acid may be titrated as a monobasic one, if an indicator is employed which changes colour with a concentration of hydrogen ions which corresponds with the concentration of hydrogen ions of the primary salt. Again a polybasic acid may be titrated as a dibasic acid if the indicator which is employed is one which changes colour with the concentration of hydrogen ions of the secondary salt, and so on. A deci-normal solution of dihydrogen sodium phosphate has been found to possess a hydrion concentration of $9 \cdot 3 \times 10^{-5}$ normal; a deci-normal solution of hydrogen disodium phosphate has been found to possess a hydrion concentration of $1 \cdot 3 \times 10^{-9}$ normal. If these numbers be compared with the dissociation constants of methyl orange and of phenolphthalein, the behaviour of phosphoric acid on titration with these two indicators becomes readily intelligible.

The question may be asked—Can orthophosphoric acid be titrated as a tribasic acid? The concentration of hydrogen ions in a solution of trisodium phosphate has been determined as $4 \cdot 3 \times 10^{-13}$ normal; so that if an indicator be selected which changes colour with a concentration of hydrogen ions corresponding to that particular number, the titration of phosphoric acid as a tribasic acid ought to be possible experimentally.

The sensitiveness of indicators

If we know the dissociation constant of an indicator, we can say within what particular concentrations of hydrogen ions its colour change will take place. This may be tested by preparing a number of aqueous solutions so that the concentrations* of hydrogen ions in them are respectively 10^{-3}, 10^{-4}, 10^{-5}, 10^{-6}, 10^{-7}, 10^{-8}, 10^{-9}, 10^{-10}, 10^{-11}. The concentration of hydrogen ions in pure water at 25° C. is approximately 10^{-7}. The range of sensitiveness of an indicator can then be tested by placing small equal quantities of it in turn in the different solutions. It will be found that methyl orange is completely red in the 10^{-3} solution, orange coloured in the 10^{-4} solution, and yellow at 10^{-5}. Methyl red is completely red at 10^{-4}, light red at 10^{-5}, and yellow at 10^{-6}. Again phenolphthalein is colourless at a concentration of hydrogen ions of 10^{-7}, faintly coloured at 10^{-8}, and deeply coloured at 10^{-9}. Reference to the table giving the dissociation constants of these indicators will show that except in the somewhat doubtful case of phenolphthalein the values of these constants lie within the limits of the range of sensitiveness.

Since the concentration of hydrogen in pure water is approximately 10^{-7} at 25° C., it follows that the exactly neutral point is only determined by an indicator which has a dissociation constant of about 10^{-7}. It does not follow, however, that an indicator with a dissociation constant of that particular value is necessarily therefore the best to use, but it may be stated generally that the most useful indicators are those with dissociation constants not greatly different from 10^{-7}. An indicator must therefore be a weak acid or a weak base, but it must not be too weak; for example an indicator with a dissociation constant of 10^{-11} which undergoes a colour change at concentrations of hydrogen ions of 10^{-10} and 10^{-12}, that is between concentrations of hydroxyl ions of 10^{-4} and 10^{-2} (since the product of the concentrations of the hydrogen and hydroxyl ions in water is 10^{-14} at 25° C.), would be too weak. A fairly considerable quantity of alkali is necessary to effect the change in such cases.

Considering now the process of titration of an acid by a base, the effect of the gradual addition of alkali to the acid is to diminish the concentration of the hydrogen ions in the solution until a point is reached when the indicator commences to dissociate electrolytically to an appreciable extent. This point may or may not be the so-called equivalent point, that is to say the point when

* In gramme ions per litre.

exactly equivalent quantities of acid and base are present together. In titration, we may either stop the addition of alkali as soon as we observe a distinct colour change or we may continue the process until further slight addition of alkali has no more appreciable effect. If we are working with a two-coloured indicator such as methyl orange or methyl red, and if we stop the titration as soon as we observe a distinct colour change, the end-point will be independent of the amount of the indicator present, because we are titrating to a certain fractional change of the indicator. This, however, does not take account of the fact that the neutralization of the indicator does require a certain amount of alkali for its completion, and obviously the more of the indicator which is present the greater will be the amount of alkali required. The difficulty does not arise, however, if the indicator is put into the solution in the same form as it will have at the end of the titration, that is to say if methyl orange or methyl red be used in the form of their sodium salts. The effect of the amount of indicator present in the solution on the end-point is somewhat different in the case of an indicator such as phenolphthalein, in which we continue the addition of alkali until there is a certain colour in the solution, that is to say until there is a certain amount of coloured substance formed. If there is a fairly large quantity of indicator present, the amount of coloured substance may be but a small fraction of the total quantity of indicator; but if the quantity of indicator is small, the amount of coloured substance may be a relatively large fraction of the total quantity of indicator. Since the equilibrium between the undissociated molecule of an indicator and its ions is regulated by the law of mass action, viz.

$K \times$ (concentration of undissociated molecule)

$=$(concentration of hydrion) $\times$ (concentration of anion),

it is clear that the greater the amount of the indicator, the more sensitive will it be to small concentrations of hydrogen ions. In the particular case of phenolphthalein, while it is true that the larger the quantity of the substance present, the more sensitive it is to small concentrations of hydrogen ions, there is a limit to its sensitiveness owing to its very limited solubility.

The manner in which the concentration of hydrogen ions varies in a solution during the process of titration may be demonstrated in a satisfactory manner by the curve in Fig. 5. This curve represents the change in the concentration of hydrogen ions when the titration of 50 c.c. of centi-normal hydrochloric acid by centi-normal sodium hydroxide is almost complete. The abscissae represent the concentration of hydrogen ions, while the ordinates represent the number of cubic centimetres of base added. When 49·95 c.c. of base are added, the concentration of hydrogen ions is 10^{-6}; when 50·05 c.c. are present, the concentration of hydrogen ions is 10^{-8}. Along the curve are written the various indicators at points which correspond to the particular end-points they indicate. It will be observed that methyl red, litmus, and phenolphthalein come on the flat point of the curve; in other words

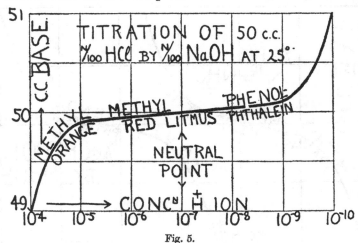

Fig. 5.

these indicators all give sharp end-points. For methyl red, the result obtained, viz. 49·95 c.c. instead of the theoretical 50 c.c., is correct to about one part in one thousand. Phenolphthalein gives an equally satisfactory end-point, the result of which gives an error of the same order of magnitude in the other direction. Methyl orange, however, occurs on the steep part of the curve, and has the additional disadvantage that its colour changes comparatively slowly. The end-point observed with this indicator will be about 49·8 c.c. giving a result considerably less accurate than the other indicators.

Very interesting results are obtained when either the acid or the base is weak. In the adjoining diagram (Fig. 6) curves are given representing respectively the titration of hydrochloric acid by ammonia and of acetic acid by sodium hydroxide, all the substances being in centi-normal solution. On account of hydrolysis, the concentration of the hydrogen ions in the solution, when acid and base are present in equivalent proportions, is not the same as in pure water. The addition of excess of acid or base does not alter the concentration of the hydrogen ions to a great extent; because one effect of such addition is to diminish the degree of hydrolysis; and, further, if the weak electrolyte is

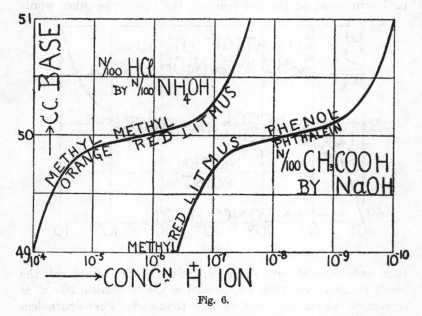

Fig. 6.

present in excess, the degree of ionization is reduced by the presence of the neutral salt. A glance at the curves will show that although in these cases the degree of hydrolysis is small and easily reduced to *nil* by the addition of a slight excess of either constituent, the flat part of the curve is very much curtailed. Instead of finding that there are a number of indicators which determine neutrality with accuracy, we find only methyl red in one case, and phenolphthalein in the other. The use of other indicators than those mentioned gives very inaccurate results, and the order of magnitude of the errors is clearly indicated by the curves.

In cases where either the acid or the base is still weaker, the flat portion of the curve becomes curtailed to a still further extent and finally disappears altogether. No indicator will give a satisfactory result in such cases, and accurate titration becomes impossible. A good example of a weak base with which accurate results cannot be obtained is aniline. Aniline hydrochloride is hydrolysed to such a great extent that the concentration of hydrogen ions in solution at the equivalent point is $10^{-3.5}$; a large excess of acid or base produces only a slight change in this

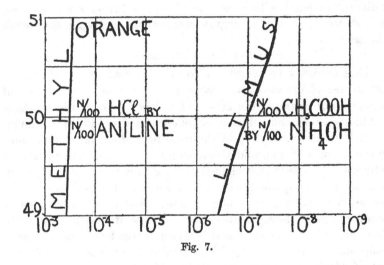

Fig. 7.

value. The titration of centi-normal aniline by hydrochloric acid of this concentration is shown by the curve in Fig. 7.

In the previous chapter it was stated that satisfactory results cannot be obtained by titrating a weak acid with a weak base. This fact is well shown by the curve representing the titration of centi-normal acetic acid by centi-normal ammonia. Since, however, there is never any necessity to titrate a weak acid by a weak base we need not discuss this matter any further.

Other theories of indicators

It must be pointed out that the ionic theory of indicators is by no means generally accepted by all chemists. It is contended by some that it fails to give a sufficient explanation of all the facts, and also that the change of colour in acid and alkaline solution may be explained as due to a change in the constitution of the indicator*. It would be beyond the scope of this work to enter into a detailed discussion of the relative merits of these theories, but it must be pointed out that the ionic theory is the only one by which the observed phenomena receive quantitative explanation. It must also be pointed out that the ionic theory by no means precludes the possibility of change of constitution of the indicator in passing from acid to alkaline solution and *vice versa*, indeed we have the strongest reasons for believing that such change occurs.

Change of constitution

It is contrary to experience to observe that simple change in the degree of ionization is accompanied by a striking change of colour. For example the colour of a solution of cupric sulphate is blue whether the solution is concentrated or dilute, although the salt is present chiefly in the form of undissociated molecules in the strong solution, and almost completely ionized in the dilute solution. It is true that change of the concentration of a solution is usually accompanied with an alteration of the *depth* of the colour, but we have abundant evidence that a radical change of colour is usually brought about by a change of constitution. We therefore regard indicators as substances of which the undissociated molecule consists of an equilibrium mixture of tautomeric forms, one of which only ionizes to any considerable extent. This conception does not affect the discussion of indicators from the point of view of electrolytic dissociation, provided we bear in mind that the undissociated molecule is really an equilibrium mixture of the tautomeric forms; indicators being regarded not as true but as pseudo acids or bases. According to this view the change in ionization is accompanied by a tautomeric change, and it is clear

---

* The discussion of the chemical constitution of the acidic and basic forms of indicators can scarcely be considered from a general point of view. It is best to confine the discussion to particular cases.

that the tautomeric change must be a rapid one in order that the indicator may determine neutrality with sharpness. Ionic reactions are practically instantaneous, but this is not necessarily the case with tautomeric reactions, and some indicators are somewhat slow in their colour change on this account.

On the simple ionic theory an indicator was represented by a formula such as XOH, which on undergoing electrolytic dissociation was split up into the ions $\overset{-}{X}O$ and $\overset{+}{H}$, the undissociated molecule XOH and the anion $\overset{-}{X}O$ being differently coloured in solution. On the modified theory according to which the indicator consists of two or more tautomeric forms in equilibrium, the indicator might be represented as an equilibrium mixture of the forms XOH and HXO, the form XOH being capable of dissociating into the ions $\overset{-}{X}O$ and $\overset{+}{H}$. The equilibrium equation might be written as

$$HXO \rightleftharpoons XOH \rightleftharpoons \overset{+}{H} + \overset{-}{X}O.$$

The form XOH and the ion $\overset{-}{X}O$ must be similarly coloured, but the form HXO must be differently coloured from these. If in the equilibrium between the two tautomeric forms there is a great excess of the form HXO, the colour of the solution will be determined by that form. However, if equilibrium be disturbed by ionization taking place, the form XOH will disappear, and a readjustment of equilibrium between the two tautomeric forms will take place. The process may go on until the solution contains a large amount of the ion $\overset{-}{X}O$ and only a very small amount of the two forms HXO and XOH; the colour of the solution will now be determined by the colour of the ion $\overset{-}{X}O$. It is clear that the general theory of indicators is not materially altered by the substitution of the equilibrium mixture of the tautomeric forms for that of the simple undissociated molecule.

REFERENCES

The Sensitiveness of Indicators. By H. T. Tizard. (*British Association Report*, 1911, p. 268.)

Quantitative Application of the Theory of Indicators to Volumetric Analysis. By A. A. Noyes. (*Journ. Amer. Chem. Soc.*, 1910, 32, p. 815.)

CHAPTER X

UNCLASSIFIED VOLUMETRIC DETERMINATIONS

In this chapter we shall give an account of methods of determining various substances by methods which cannot readily be classified under any general scheme, although some of them are concerned with oxidation and precipitation reactions. Some of the methods are empirical in the sense that they require approximately uniform conditions of standardizing the reagent and applying it to the particular determination which it is desired to carry out.

Determination of copper by the cyanide method

When a solution of ammonia is added to a solution of a cupric salt, a precipitate of cupric hydroxide is first formed, which readily dissolves in excess of ammonia forming a deep blue solution of a complex copper ammonium salt. If potassium cyanide be added to the ammoniacal copper solution, the blue colour is gradually discharged until it ultimately disappears*. Upon these reactions a rapid method of determining copper has been based, but the method is a purely empirical one, as the quantity of potassium cyanide required to decolorize a solution containing a given quantity of copper depends to a considerable extent upon the amount of free ammonia in the liquid and upon other factors. It is therefore not advisable to attempt to represent the reactions which occur by means of equations, but to standardize the solution of potassium cyanide with reference to pure electrolytic copper,

---

* The reactions which occur are very complicated and it is perhaps scarcely advisable to attempt to represent them by equations.

and to adhere as closely as possible in carrying out the actual determinations to the conditions of experiment observed in standardizing the cyanide solution.

The standard copper solution is prepared by dissolving a known weight of pure electrolytic copper (about 5 grammes) in nitric acid diluted with its own volume of water, care being taken to guard against loss of liquid by effervescence. After solution of the metal, it is advisable to add a small quantity of bromine water and to boil the liquid till the bromine is removed. The object of this treatment with bromine water is to oxidize any nitrous acid formed by the reduction of the nitric acid by the metal. The solution is boiled until the greater part of the free acid has been expelled by evaporation, and is then diluted with water and made up to a known volume.

A solution of potassium cyanide containing about 30 grammes of the salt per litre is next prepared, and standardized upon the copper solution in the following manner. A measured quantity of the copper solution is taken and ammonia added in slight excess of the amount required to produce a dark blue solution, the amount of ammonia solution being noted. The potassium cyanide solution is then added from a burette until a very pale lavender tint remains in the liquid. It is better to stop the addition of the cyanide solution at this point rather than to continue it until the colour is totally discharged, as the end-point is more easily observed in the former case. The titration is then repeated a second and a third time, using the same amount of ammonia solution in each case.

When the copper value of the potassium cyanide solution has been obtained, the determination of copper in an unknown solution may be made. It is however very important to adjust the conditions so as to be approximately identical in each case; otherwise the method will not give accurate results. The potassium cyanide solution should be standardized at frequent short intervals, as the salt is unstable in aqueous solution, becoming hydrolysed to formate and other products. It has been stated by some operators that more uniform results are obtained if the solution is made alkaline with some other substance than ammonia, and ammonium carbonate has been recommended for this purpose.

Determination of ferric iron by reduction with titanous chloride solution

An aqueous solution of titanous chloride $TiCl_3$ possesses powerful reducing properties, becoming itself oxidized to titanic chloride $TiCl_4$. Thus a solution of ferric chloride is quantitatively reduced to ferrous chloride according to the equation

$$FeCl_3 + TiCl_3 = FeCl_2 + TiCl_4.$$

The solution of titanous chloride, which possesses a beautiful purple colour, must be preserved in an atmosphere of hydrogen on account of the great ease with which it undergoes oxidation to titanic chloride which is colourless. The end-point of the reduction of the ferric salt is readily observed by the addition of some ammonium thiocyanate solution, and the addition of the titanous chloride solution is continued until the dark red solution becomes colourless.

Titanous chloride in concentrated solution is now a commercial product, and the standard solution is prepared by boiling about 50 c.c. of the commercial solution with about 100 c.c. of strong hydrochloric acid. The mixture is then diluted with water to about 2·25 litres and preserved in a reservoir which is in connexion with an apparatus for generating hydrogen from zinc and hydrochloric acid. The burette for delivering the liquid is placed in direct connexion with the reservoir.

The iron value of the titanous chloride solution is obtained by dissolving a known weight of pure ferrous ammonium sulphate in water with the addition of a little sulphuric acid and diluting the mixture to a suitable volume. A solution containing about four grammes of the double salt in 200 c.c. will be found of a suitable strength. An aliquot portion of this solution (20 c.c.) is then withdrawn and dilute potassium permanganate solution is added until a faint pink colour remains in the liquid. A moderate excess of ammonium thiocyanate solution is then added, and the titanous chloride solution is then run in from the burette until the red colour of the solution vanishes. Further titrations are carried out in the same way, and the iron value of the titanous chloride solution determined. The double iron salt contains one-seventh of its own weight of iron. It is clear that it is unnecessary to know the strength of the permanganate solution. The titanous chloride solution can then be employed for the direct determination of an unknown solution of a ferric salt. It is immaterial whether the iron be present in the form of chloride or sulphate, but the presence of free mineral acid is necessary, as otherwise the end-point of the reaction will not be clearly defined. The titanous chloride solution should be titrated against a standard solution of iron at frequent short intervals, even although protected from atmospheric oxygen by being preserved in a hydrogen atmosphere*.

Determination of the available chlorine in bleaching powder by standard arsenite solution

The chlorine which is present in bleaching powder in the form of hypochlorite when the substance is treated with water may be readily and accurately determined by oxidation of a solution of an alkaline arsenite. The arsenious oxide is converted by the available chlorine into arsenic oxide, 198 parts by weight of arsenious oxide being equivalent to 4×35.5 parts by weight of available chlorine. The method of preparing a standard solution of sodium arsenite has been described in Chapter IV. A suitable quantity of the sample of bleaching powder is weighed out, triturated with water, and made up to a known volume as described in Chapter V. Aliquot portions of the milky liquid are then withdrawn and the standard solution of sodium arsenite allowed to flow in from the burette until the reaction is completed.
end-point of the reaction is determined by means of starch potassium iodide paper employed as an external indicator. This paper is prepared by soaking pieces of filter paper in a solution of starch to which a little potassium iodide has been added. After drying, the paper will be found a sensitive test for certain oxidizing agents such as chlorine. In using starch potassium iodide paper in the determination of bleaching powder, drops of the liquid undergoing titration are removed by means of a glass rod and brought in contact with the paper. As long as any available

---

* For further applications of titanous chloride the reader is referred to the work of Knecht and Hibbert, *New Reduction Methods in Volumetric Analysis*, 1910.

chlorine remains in the liquid, the paper will be stained a dark colour, but as soon as the reaction has been completed, the paper will remain colourless.

Instead of carrying out the determination with the aid of an external indicator, some chemists prefer to add a measured quantity of the standard solution of sodium arsenite, which must be in excess of the amount required to react with the quantity of bleaching powder taken. The excess of alkaline arsenite is then determined by back titration with standard iodine.

Determination of zinc by means of potassium ferrocyanide

When a solution of potassium ferrocyanide is added to a solution of a zinc salt a sparingly soluble double ferrocyanide of potassium and zinc is precipitated. Upon this reaction a method of determining zinc has been based. This method is of an empirical nature and consequently the conditions of experiment should be adjusted so as to be approximately identical in the different determinations. The ferrocyanide solution is prepared by dissolving about 35 grammes of the crystallized salt in water and diluting the solution to one litre. The solution is standardized by means of a solution of zinc sulphate or chloride prepared by dissolving about one gramme of pure zinc in hot dilute sulphuric or hydrochloric acid, care being taken to avoid loss of liquid by effervescence. The solution of the metal is diluted with water and made up to a known volume (200 c.c.). A large excess of free acid should be avoided, but the solution should certainly contain some free acid. Aliquot portions of the solution are then withdrawn and heated to about 90° C. and the solution of potassium ferrocyanide added from a burette until a drop of the liquid when brought in contact with a drop of a solution of uranyl acetate on a spot plate shows a brown colour due to the formation of uranyl ferrocyanide, indicating that a slight excess of ferrocyanide has been added. The titration should not be carried out too rapidly, since the precipitation of the double ferrocyanide of zinc and potassium takes place somewhat slowly. The end-point with uranyl acetate is usually clearly defined in the absence of certain metals such as iron. When care is taken, it is not difficult to obtain concordant results with this method. The zinc value of the ferro-cyanide solution having been obtained, the standard solution may be employed to determine unknown zinc solutions. Some chemists employ ammonium molybdate instead of uranyl acetate as an external indicator.

Determination of zinc by means of sodium sulphide solution

Zinc is readily precipitated as sulphide in an alkaline solution, and an excellent volumetric method of determining this element has been based upon this reaction. The standard solution of sodium sulphide is prepared by saturating a strong solution of sodium hydroxide with hydrogen sulphide, and adding more caustic soda until the odour of the gas has been removed. The reactions which occur may be represented by the equations

$$NaOH + H_2S = NaHS + H_2O,$$

and
$$NaHS + NaOH = Na_2S + H_2O.$$

The solution is standardized by means of a solution of a zinc salt prepared by dissolving a known weight of the pure metal in dilute hydrochloric acid and diluting the solution to a suitable volume. The solution of sodium sulphide is titrated against the zinc solution in the following manner. An aliquot portion of the zinc solution is measured out, and a mixture of ammonia and ammonium carbonate added in sufficient quantity to redissolve the precipitate which first forms. The sodium sulphide solution is added from a burette, and drops of the liquid undergoing titration are brought in contact with drops of sodium nitroprusside solution on a spot plate until the presence of a trace of alkaline sulphide in excess is shown by the appearance of a beautiful violet colour with the nitroprusside indicator. The reaction that takes place may be represented by the equation

$$ZnSO_4 + Na_2S = ZnS + Na_2SO_4.$$

The solution of sodium sulphide should then be diluted to a suitable strength for the determinations. It is best to ascertain the zinc value of the standard solution and employ that value rather than to attempt to calculate the strength of the solution from the chemical equation. Some chemists prefer to use an alkaline solution of lead tartrate instead of sodium nitroprusside as the external indicator, the end-point of the reaction being indicated by the formation of a black precipitate of lead sulphide. Nickelous chloride has also been employed for the same purpose.

Determination of formaldehyde

Formaldehyde, $H.CHO$, is a gas which is readily soluble in water. The product containing about 40 per cent. of formaldehyde is known commercially as formalin, and is much used as a disinfectant. The substance in dilute aqueous solution may be determined by a number of methods, one of the best depending upon the oxidation of the aldehyde by means of iodine in alkaline solution. When an aqueous solution of formaldehyde is oxidized by means of iodine in presence of sodium hydroxide, sodium formate and iodide are formed according to the equation

$$H.CHO + I_2 + 3NaOH = H.COONa + 2NaI + 2H_2O.$$

In carrying out the determination, the measured quantity of the dilute aqueous solution of formaldehyde is mixed with a measured quantity of standard iodine solution, which must be in excess. Sodium hydroxide solution is then added until the liquid becomes of a pale yellow colour. The mixture is allowed to react for ten minutes, and then the solution is acidified with dilute hydrochloric acid to liberate the excess of iodine. This liberated iodine is then determined by titration with standard sodium thiosulphate solution. The iodine which has been used up in the oxidation of the formaldehyde is thus determined by difference.

Determination of phosphates

The determination of phosphates may be effected volumetrically by a number of methods, and for general purposes it is very difficult to recommend any one method in preference to another. For ordinary work, the uranium method is perhaps the simplest, and with care it can be made to yield satisfactory results. This method is, however, an empirical one, and consequently it is important to see that the same conditions of experiment are observed in standardizing the uranium solution as are employed in carrying out the determination of phosphates in unknown solutions.

The determination of phosphates by precipitation with uranyl nitrate or acetate depends upon the fact that the phosphate radical is precipitated as uranyl phosphate. The precipitation takes place somewhat slowly in the cold, but more rapidly at a higher temperature; a temperature of about 90° C. yields good results. The end-point of the reaction is determined by means of potassium ferrocyanide solution on a spot plate, when the presence of the slightest excess of the uranium solution is shown by the appearance of the brown colour of uranyl ferrocyanide. If uranyl nitrate is employed as the standard solution, it is necessary to add sodium acetate to the liquid undergoing titration in order to prevent the possibility of the occurrence of free nitric acid in the solution; with uranyl acetate, the addition of sodium acetate is unnecessary, but if added it will do no harm.

The standardization of the uranium solution may be effected either on pure tricalcium phosphate $Ca_3(PO_4)_2$ or on hydrogen sodium ammonium phosphate $NaNH_4HPO_44H_2O$ (microcosmic salt). The former salt is to be preferred if the uranium solution is required for the determination of phosphoric acid in combination with calcium or magnesium. A suitable quantity, say 10 grammes, of tricalcium phosphate is weighed out and dissolved in a little nitric acid, and the liquid diluted to a litre. If there is any doubt about the purity of the calcium phosphate, the phosphorus should be determined gravimetrically as magnesium pyrophosphate.

The uranium solution is prepared by dissolving about 35 grammes of either the acetate or the nitrate in water and diluting the solution to one litre. This solution is standardized by measuring out a suitable volume of the standard phosphate solution, heating to about 90° C., and titrating with the uranium solution until a drop of the liquid when brought in contact with a drop of potassium ferrocyanide on a spot plate produces a permanent brown colour. If the titration is made with uranyl nitrate, a measured quantity of a solution of sodium acetate containing acetic acid must be added in each titration; about 5 c.c. of a solution containing 100 grammes of sodium acetate and 50 c.c. of glacial acetic acid should be added in each titration. The phosphate value of the uranium solution is then determined. For most purposes, it will be found convenient to express the strength of the uranium solution in terms of its equivalence to phosphorus pentoxide P_2O_5.

The uranium method of determining phosphates requires considerable care, especially as regards determining the end-point by means of the ferrocyanide indicator. The precipitation of uranium phosphate takes place somewhat slowly, with the result that a colour reaction with potassium ferrocyanide is sometimes observed before precipitation is complete, if the titration is carried out too hastily. A little practice will soon enable the operator to ascertain when the true end-point has been reached.

Determination of sodium

Sodium is a metal which forms very few insoluble salts, and consequently the precipitation of this element from solution can only be effected in a few isolated cases. It has been shown however that the sodium salt of dihydroxytartaric acid is a very sparingly soluble substance, and, since this acid may be oxidized to carbon dioxide and water by means of potassium permanganate in presence of dilute sulphuric acid, a valuable volumetric method of determining sodium has been placed in the hands of chemists. For details of the experimental procedure, the student is recommended to consult the original paper by Fenton (*Trans. Chem. Soc.*, 1898, p. 167); the principles upon which the method depends however may be briefly stated.

When tartaric acid in concentrated aqueous solution is treated with hydrogen peroxide in presence of a small quantity of a ferrous salt it undergoes oxidation to dihydroxymaleic acid. The oxidation of the tartaric acid may be represented by the equation

$$\begin{array}{ccc} \text{CH(OH)COOH} & & \text{(HO)CCOOH} \\ | & + \text{O} = & \| & + \text{H}_2\text{O}. \\ \text{CH(OH)COOH} & & \text{(HO)CCOOH} \end{array}$$

The oxidation of the tartaric acid must be effected in ice cold solution, and the isolation of the dihydroxymaleic acid is carried out by addition of fuming sulphuric acid in very small quantities at a time. The dihydroxymaleic acid crystallizes with two molecules of water.

The next step is the conversion of the dihydroxymaleic acid into dihydroxytartaric acid by oxidation by means of bromine. The dihydroxymaleic acid is suspended in glacial acetic acid, and a slight excess of bromine added in small quantities at a time. A drop of water is occasionally added to the mixture. The reaction which takes place is represented by the equation

$$\begin{array}{ll} (OH)CCOOH \\ \parallel \qquad\qquad + 2H_2O + Br_2 = \\ (OH)CCOOH \end{array} \begin{array}{l} C(OH)_2COOH \\ \mid \qquad\qquad\quad + 2HBr. \\ C(OH)_2COOH \end{array}$$

After standing for some time, the dihydroxytartaric acid is precipitated as a heavy crystalline solid. The product is purified by washing with anhydrous ether after filtering off under pressure.

Fenton showed that when this acid is treated with potassium permanganate in presence of sulphuric acid it undergoes complete oxidation to carbon dioxide and water. It would appear that one molecule of dihydroxytartaric acid should theoretically require three atoms of oxygen for complete oxidation according to the equation

$$C_4H_6O_8 + 3O = 4CO_2 + 3H_2O.$$

But experiment shows that the amount of oxygen required is always less than the theoretical amount. The actual quantity of oxygen required corresponds very nearly to 2·9 atoms. It is not clear to what the difference between the experimental and the theoretical values is due, although it may be due to the slow decomposition of the dihydroxytartaric acid in aqueous solution into tartronic acid and carbon dioxide. Tartronic acid is oxidized by permanganate, but with much greater slowness than dihydroxytartaric acid. The action of potassium permanganate on dihydroxytartaric acid in presence of dilute sulphuric acid is slow at first, but soon becomes much more rapid and finally becomes slow again as the end of the reaction is approached. The end-point is however quite definite.

In carrying out a determination of sodium, a suitable quantity of dihydroxytartaric acid is exactly converted into the potassium salt by neutralization with potassium carbonate. The neutralization is effected by dissolving the acid and potassium carbonate in equivalent proportions in separate small quantities of ice cold water, and mixing the solutions. The potassium salt separates in the crystalline form. A known weight of pure sodium chloride is then taken, dissolved in the minimum quantity of water, and cooled in ice. A considerable excess of potassium dihydroxy-tartrate is also dissolved in the least quantity of ice cold water; the two solutions are mixed and kept at a low temperature for some time. Sodium dihydroxytartrate is precipitated by double decomposition. The precipitate is collected on a filter, washed with a small quantity of ice cold water, and dissolved in excess of dilute sulphuric acid. The resulting solution is then titrated with potassium permanganate.

It will be observed that the permanganate is standardized with reference to pure sodium chloride. Having determined the sodium value of the permanganate solution the determination of sodium in other compounds may be carried out. It is better to proceed in this way rather than to calculate the result from the fact that one molecule of dihydroxytartaric acid is oxidized by approximately 2·9 atoms of oxygen.

Determination of the hardness of waters

Natural waters are classified as hard or soft according as they require much or little soap to make a lather. The hardness of many natural waters is due to the presence of calcium or magnesium salts in solution. When a hard water is brought in contact with a soap, a substance which consists of the sodium or potassium salts of certain higher fatty acids, double decomposition takes place with the formation of a precipitate which consists of the calcium or magnesium salts of the fatty acids derived from the soap. The hardness of a water may be temporary, that is, the water is capable of being softened by boiling. Temporary hardness is due to the presence of calcium carbonate, which is held in solution by carbon dioxide in the form of a soluble bicarbonate. When the water is boiled, the carbon dioxide is expelled from solution resulting in the precipitation of the calcium carbonate. The cause of the permanent hardness of a water, that is, of the

hardness which cannot be removed by boiling, is the presence of
dissolved calcium or magnesium sulphate.

A method of determining the hardness of waters was devised
many years ago by Clark. This method consists in adding a
standard solution of soap to a measured volume of the water with
frequent shaking until a permanent lather is obtained. The first
portions of the soap are used up in the precipitation of the insoluble
calcium and other salts, but as soon as excess of soap has been added,
a permanent lather is obtained. The soap solution is prepared by
dissolving a suitable quantity of Castille soap in dilute alcohol. The
solution is made of such a strength that 1 c.c. of it will precipitate
exactly 1 milligramme of calcium carbonate in solution. The soap
solution is standardized by weighing out one gramme of Iceland
spar, dissolving this in excess of dilute hydrochloric acid, and
evaporating the solution to dryness. The residue is dissolved in
distilled water and diluted to one litre. Measured quantities of
this solution of calcium chloride are taken, and the standard soap
solution added from a burette until a permanent lather is obtained.
It is important always to conduct the titrations in the same manner.
The titrations are carried out in stoppered bottles in order to
permit vigorous shaking. The soap solution should be added in
small quantities at a time, shaking carefully between each addition.
Even when the quantity of soap which is approximately required
is known, the reagent should be added in the manner described.
The end-point is taken as that point at which the contents of the
bottle possess a permanent unbroken lather over the surface of
the liquid. An experiment is also made to ascertain how much
soap solution is required to produce a permanent lather with a
certain volume of distilled water. This correction is applied to
the result obtained in standardizing the soap solution. In determining the hardness of a water, a suitable volume of it is titrated
in the manner already described. The results are usually expressed
in so-called degrees of hardness or parts by weight of calcium
salts or their equivalent in 70,000 parts of water. This scale
corresponds to parts by weight in grains of calcium carbonate to
a gallon of water. To simplify the arithmetical work it is usual
to titrate 70 c.c. of the water at a time; if the soap solution is
properly standardized, the degrees of hardness are obtained

directly by reference to a table. The use of a table is necessary, since the volume of soap solution which must be added to produce a permanent lather is not strictly proportional to the amount of calcium salt present. That is to say, if 1 c.c. of the soap solution will precipitate 1 milligramme of calcium carbonate, a solution containing n milligrammes of calcium carbonate will not require n c.c. of soap solution but a smaller volume. This departure from direct proportionality plainly indicates that the theory of the process is much more complicated than that which has been given; nevertheless the process gives satisfactory results when properly carried out.

The process which has been described determines the total hardness, that is, the sum of the temporary and permanent hardnesses. To determine the permanent hardness alone, a volume of the water equal to that originally taken for the determination of the total hardness is boiled for about half an hour and made up to its original volume with distilled water. This water is then titrated with standard soap solution in the usual way. The result gives the permanent hardness, and the difference from the total hardness gives the temporary hardness.

A preferable method of determining the hardness of water was devised by Hehner. The temporary hardness of the water is determined by direct titration with standard sulphuric acid. For this purpose $\frac{N}{50}$ sulphuric acid is used. One cubic centimetre of acid of this strength will neutralize exactly one milligramme of calcium carbonate. In carrying out a determination of the temporary hardness 100 c.c. or 70 c.c. of the water are heated nearly to boiling, a small quantity of a suitable indicator being added to the liquid. Very good results are obtained with phenacetolin. This indicator is pink in solutions of alkaline carbonates, but golden yellow in acid solution. The addition of the standard acid is continued until the correct colour change takes place. The result gives the degree of temporary hardness directly in parts per 70,000 parts of water.

The permanent hardness is obtained by double decomposition with a standard solution $\left(\frac{N}{50}\right)$ of sodium carbonate. Each cubic centimetre of this solution will precipitate one milligramme of calcium carbonate or its equivalent in magnesium salts from solution. A measured volume of the water to be examined (70 c.c.) is taken, and a suitable known excess of standard sodium carbonate added. The solution is evaporated to dryness in a platinum dish, the soluble portion extracted with distilled water, filtered, and the filtrate titrated with standard sulphuric acid. The titration represents the amount of sodium carbonate added in excess; the difference representing the permanent hardness.

If the water contains alkali carbonate, the apparent temporary hardness as determined by titration with acid will be greater than the true value. The determination of the permanent hardness on the other hand may show more sodium carbonate than was actually added. In such a case there is no permanent hardness since the salts to which the hardness is due are decomposed by the alkali carbonate. The true temporary hardness is obtained by deducting the apparent increase in the amount of sodium carbonate added from the temporary hardness as determined by the acid titration; the result will be the true temporary hardness.

CHAPTER XI

SOME APPLICATIONS OF VOLUMETRIC METHODS

The reader who has worked through the previous chapters of this book will have become aware of the possibilities of combining two or more different volumetric processes together in order to determine the constituents of various mixtures. The availability of volumetric methods for work of this kind is very great, and it is the object of the present chapter to illustrate a few of the determinations which may be effected in this way. In many cases, alternative methods to those which are suggested are available.

(1) *Determination of oxalic and sulphuric acids when present together in the same solution*

The oxalic acid is determined by titrating a measured portion of the solution by standard potassium permanganate, the solution being as usual warmed to increase the velocity of the reaction. If the quantity of sulphuric acid present in the mixture is insufficient to prevent the precipitation of hydrated manganese dioxide, fresh dilute sulphuric acid must be added in each titration. Then in another experiment, measured portions of the solution are titrated with standard sodium hydroxide, phenolphthalein being employed as an indicator. From the titration with caustic soda we determine the quantity of soda required to neutralize both acids. Then, knowing the amount of oxalic acid which is present in a given volume of the solution from the permanganate titration, we can calculate how much sodium hydroxide has been employed in the neutralization of the oxalic acid alone. The difference between the amount of caustic soda found by experiment and that which has been employed in the neutralization of the oxalic acid is clearly equal to that which has been employed in the neutralization of the sulphuric acid, which is thus determined.

(2) *Determination of the amounts of ammonium chloride*
and ammonium sulphate in a mixture

The ammonium chloride is determined by direct titration with standard silver nitrate in the usual way, potassium chromate being employed as indicator. Then another portion of the solution is boiled with a measured excess of standard sodium hydroxide until all ammonia is expelled from the solution. The amount of caustic soda remaining in excess is then determined by titration with standard acid. The difference between the amount of caustic alkali originally taken and that determined by titration is clearly equal to the amount required to decompose the ammonium chloride and sulphate in the portion of the mixture taken. In this way, the total quantity of the radical ammonium, NH_4, is determined. Knowing the amount of ammonium chloride from the titration with silver nitrate we can calculate the amount of ammonium sulphate by difference.

(3) *Determination of the amounts of sodium chloride*
and sodium hydroxide in a solution

The sodium hydroxide is determined first by titration with standard nitric or sulphuric acid. Then to the neutral solution thus obtained, two drops of potassium chromate are added, and the sodium chloride determined by titration with standard silver nitrate. It is best, after determining the sodium hydroxide, to neutralize another portion of the solution exactly without the use of any indicator, and to employ the solution thus prepared to estimate the chloride, as the colour change of the chromate indicator is more easily seen in the absence of litmus or other acidimetric indicators.

(4) *Determination of the amounts of hydrochloric acid*
and sodium chloride in a solution

The hydrochloric acid is determined by direct titration with caustic soda. The total chloride is then determined by titrating the neutral solution thus obtained with standard silver nitrate. As in the last example it is best to prepare an exactly neutral solution for determination of the chlorine without any acidimetric indicator. From the silver nitrate titration, we determine the total amount of chloride; the amount of chlorine combined with hydrogen is calculated from the titration with caustic alkali. The chlorine combined with sodium is then calculated by difference.

(5) *Determination of the amounts of ammonium chloride and sodium hydroxide in a solution*

If a solution of these two constituents be prepared, the amounts of the constituents may be determined by first titrating the sodium hydroxide with standard acid using methyl orange as indicator, and then decomposing the ammonium chloride by boiling another portion of the solution till free from ammonia, and determining by titration with standard acid the amount of sodium hydroxide remaining in excess. This procedure will clearly be successful only if the amount of sodium hydroxide in the original mixture is in excess of the amount required to decompose the ammonium salt. If the solution contains the two constituents in such proportions that the amount of ammonium chloride is not present in sufficient quantity to effect the complete decomposition of the ammonium salt by boiling the solution, a known quantity of standard sodium hydroxide must be added to the solution before boiling. In calculating the amount of ammonium chloride in this latter case, due allowance must be made for the extra quantity of sodium hydroxide added to the mixture.

This determination might also be carried out by first determining the sodium hydroxide by neutralization with standard sulphuric or nitric acid, and then titrating the ammonium chloride with standard silver nitrate in the ordinary way. Or if the method of decomposing the ammonium salt by caustic soda be employed, the procedure might be varied by passing the evolved ammonia into a measured excess of standard acid, and then determining by titration with standard alkali the amount of acid which remains unneutralized.

A solution of a mixture of caustic soda and ammonium chloride will not keep for any length of time, as the decomposition of the ammonium salt by the action of the caustic alkali takes place even at the ordinary temperature.

Further applications

The few examples that have been described will serve to give some idea of the large variety of exercises of the kind that can be devised. Many instructive experiments may be performed by estimating the same substance in different ways. For example, the strength of a solution of ferric chloride might be determined by titration with standard potassium dichromate after reduction with stannous chloride, the excess of stannous chloride being precipitated by mercuric chloride. The same substance could also be determined by the addition of a measured excess of standard silver nitrate, the amount of silver remaining in excess being determined with standard ammonium thiocyanate.

The direct determination of substances in solution, while being the main purpose of volumetric analysis, is, however, only one of the uses of this branch of practical chemistry. Many physico-chemical determinations are effected by volumetric methods, of which mention must be made of solubilities, partition coefficients, and velocity constants of chemical reactions. We shall now give a very brief account of the methods of applying volumetric analysis to problems of this kind.

Determination of solubilities

(a) *Of solids.* The first thing to be done in determining the solubility of a solid in water is to prepare a saturated solution. This may be done in two different ways. Either a quantity of the finely powdered solid is agitated with the solvent for a considerable time, the solvent being kept at a constant temperature; or advantage is taken of the fact that the solubility of most solids is greater at high temperatures than at lower temperatures, a saturated solution at the desired temperature being prepared by cooling a solution from a higher temperature. The latter method of preparing a saturated solution usually leads to higher results than the former, since the time taken for the establishment of equilibrium between solvent and solute at the ordinary temperature is very great. For most purposes it is best to prepare the saturated solution by both methods, as the true value of the solubility must clearly lie between the results obtained by the two methods. In all cases, it is essential to have excess of the solute in contact with the solution in order that the condition of saturation may be realized.

The strength of a saturated solution may be expressed in various ways, but the two most important are, first, to express the strength by stating that the saturated solution contains so many grammes of solute per litre, and second, by stating that a given volume of the solvent will dissolve a certain amount of solid. In both cases it is essential that the temperature be stated.

When a saturated solution has been prepared a suitable quantity of it is weighed out. Let the weight of saturated solution taken be w_1. In many cases, such a solution is much too strong to be titrated directly, it is therefore diluted to a measured extent. Aliquot portions of the diluted solution are then withdrawn and the strength determined by titration. The weight of solute in the weight of saturated solution w_1 is then calculated. Let the weight of solute thus determined be w_2. Then the quantity of water which has dissolved this weight of solute is clearly $w_1 - w_2$.

Among the various substances which are suitable for exercises in the determination of solubility may be mentioned oxalic acid, potassium dichromate, and ammonium chloride.

(b) *Of gases.* In discussing the solubility of gases in liquids, we have two main types of gas to deal with; those which obey the law of Henry, and those which do not. The former gases are the sparingly soluble ones, the solubility of which is best determined by absorptiometer methods. The latter gases are highly soluble in water and include gases such as ammonia and hydrogen chloride the properties of which lend themselves very well to solubility determination by volumetric methods. The preparation of saturated solutions of these very soluble gases is best effected by the use of a narrow U-shaped glass tube with a bulb near the bend. The bulb tube is first weighed empty. Then the bulb is about three-quarters filled with water, and the apparatus placed in a bath at a constant temperature.

The liquid in the bulb is then saturated with the gas. When the absorption of gas has been judged to be complete, the two ends of the bulb tube are sealed off. The bulb with its contents, and of course the ends of the tube which have been removed in the sealing, are then carefully weighed. The known weight of saturated solution thus obtained is then analysed by breaking the bulb under a suitable liquid, and the resulting solution titrated. In the case of ammonia, the bulb is broken under a measured excess of standard acid, and the excess determined by titration with standard alkali, methyl orange or methyl red being employed as indicator. In the case of hydrogen chloride, the bulb may be broken under a large excess of water, and the acid determined by direct titration with caustic soda.

Determination of partition coefficients

When a substance which is soluble in each of two immiscible solvents is shaken up with them, it distributes itself between the two solvents in a particular way. If the substance possesses the same molecular weight in both solvents, then the following simple relation holds

$$\frac{c_1}{c_2} = k,$$

where c_1 and c_2 denote respectively the concentration of the substance in the first and in the second solvent, and k is a constant.

If, however, the molecular condition of the substance is different in the two solvents, the relation is a little more complicated. If the substance associates to form a complex which possesses a molecular weight which is n times as great in the second solvent to what it is in the first solvent, then we have the relation

$$\frac{c_1}{\sqrt[n]{c_2}} = k',$$

which is an immediate consequence of the law of mass action.

The constant, which is termed the ratio of distribution or the partition coefficient of the substance between the two solvents, may be determined by shaking up the solute with measured volumes of the solvents, separating the solutions thus obtained, and determining by titration the concentration of the substance in the two solutions. Experiments are made with different total concentrations, and also with varying volumes of the two solvents. After a few determinations, it will soon be observed if the ratio of the concentrations is constant or if a more complex law is obeyed. For practice in the determination of partition coefficients, experiments may be made on the distribution of succinic acid between ether and water, and on the distribution of benzoic acid between benzene and water. In both cases, the titrations should be carried out with baryta water, phenolphthalein being used as indicator.

Determination of the velocity of chemical reactions

For an account of the theory of velocity of chemical change reference must be made to some text-book of physical chemistry. All that will be attempted here will be to illustrate the application of volumetric analysis to the determination of some substance which is taking part in a chemical change. In this connexion it may be pointed out that the usual object which the chemist has in view in determining velocity constants is to determine the so-called "order" of the reaction. While it is true that the order of a reaction agrees in many cases with the actual number of molecules which are represented as taking part in the reaction by the chemical equation, it is an undoubted fact that in a large number of cases the order of the reaction is found to be less than the number of molecules which the chemical equation represents. While this does not demand an immediate revision of the theory of the subject, it certainly suggests the necessity of a less stringent application of the order of a reaction as determined by measurement of the reaction velocity to the estimation of the number of molecules which take part in a chemical reaction.

All experiments on velocity of reaction must be performed under conditions which maintain the reacting substances at a constant temperature, since the temperature coefficient of velocity is very great. In many cases a rise of 10° C. doubles or trebles the reaction velocity.

A reaction which may be studied with success by an acidimetric method is the decomposition of dibromosuccinic acid into bromomaleic acid and hydrobromic acid. This reaction takes place in aqueous solution according to the equation

$$C_2H_2Br_2(COOH)_2 = C_2HBr(COOH)_2 + HBr,$$

and proceeds at a velocity suitable for experimental determination at a temperature of 100° C. It is clear that the solution becomes more and more acid as the reaction proceeds; the velocity of the reaction may therefore be measured by withdrawing portions of the solution at definite intervals of time and determining the acidity by titration with standard alkali.

It will be found that the rate of disappearance of the dibromo-succinic acid is at every instant proportional to the amount of this substance present in the solution or as represented by the differential equation

$$-\frac{dC}{dt} = kC,$$

where C denotes the concentration of the dibromosuccinic acid, t the time, and k the velocity constant of the reaction. The reaction is therefore a unimolecular one or as it is frequently termed a reaction of the first order. In this case, the order of the reaction is a true measure of the number of molecules taking part in the reaction.

The few examples which have been quoted will serve to convey some idea of the wide applications of volumetric work. For more detailed application we must refer the reader to special treatises.

CHAPTER XII

SOME EXAMPLES OF VOLUMETRIC DETERMINATIONS

The following examples of volumetric determinations have been given chiefly with the view of illustrating some of the inter-relationships between different processes, and also of demonstrating the order of magnitude of the errors involved. The determinations were all carried out with ordinary measuring vessels which had not been calibrated: if accurately calibrated vessels had been employed, the results would, of course, have been considerably more accurate.

(1) *Determination of the strength of a solution of potassium dichromate by means of ferrous ammonium sulphate*

A quantity of the double iron salt was purified by recrystalli-zation from hot water, a few drops of dilute sulphuric acid being added to prevent the formation of basic salt. After drying the salt by pressing between filter paper, a standard solution was prepared, and aliquot portions of this solution titrated by means of the given solution of potassium dichromate, potassium ferricyanide being as usual employed as an external indicator. The following results were obtained:

Weight of ferrous ammonium sulphate taken = 5·950 grammes.

This salt was dissolved in water with the addition of dilute sulphuric acid, and the solution diluted to 200 c.c.

Two titrations were made with 20 c.c. of the iron solution; in both cases the volume of potassium dichromate required was 15·0 c.c.

From this it follows that the volume of the dichromate solution which would be required for the complete oxidation of the 5·950 grms. of the double iron salt (corresponding to 0·850 grm. of iron) = 150 c.c.

The strength of the solution of potassium dichromate is calculated from the equation

$$\frac{150x}{0.850} = \frac{294}{6 \times 56},$$

from which $x = 0.00496$ grm. $K_2Cr_2O_7$ per c.c.

(2) *Determination of the amount of arsenious oxide in a solution of sodium arsenite by means of pure arsenious oxide*

This determination was carried out by carefully purifying some arsenious oxide by sublimation, preparing a standard solution of sodium arsenite from this, and titrating first the standard solution of alkali arsenite and then the unknown solution of sodium arsenite with a solution of iodine of unknown strength. The titrations were carried out in the usual manner with the addition of excess of sodium bicarbonate to the arsenic solutions, and the end-point determined with the aid of starch.   The following results were obtained :

Weight of resublimed arsenious oxide taken = 0·898 grm.

This solid was dissolved in a solution of sodium carbonate containing 4 grms. of the solid carbonate, and the liquid diluted to 200 c.c.

The standard solution of sodium arsenite prepared as above described was titrated with a solution of iodine of unknown strength.   Two titrations with 20 c.c. of the arsenic solution required in each case 29·95 c.c. of iodine.

The given solution of sodium arsenite was next titrated with the same iodine solution.

Two titrations with 20 c.c. of the solution in each case required 32·1 c.c. of iodine.

The calculation of the strength of the given solution of sodium arsenite is made from the equation :

$$\frac{\text{Concentration of As}_2O_3 \text{ in the given solution}}{\text{Concentration of As}_2O_3 \text{ in the standard solution}} = \frac{32 \cdot 1}{29 \cdot 95}.$$

The standard solution contained 0·00449 grm. of As_2O_3 per c.c.

Therefore the weight of As_2O_3 in each c.c. of the given solution of sodium arsenite

$$= \frac{32 \cdot 1 \times 0 \cdot 00449}{29 \cdot 95} \text{ grm.}$$

$$= 0 \cdot 00481 \text{ grm.}$$

(3) *Determination of the strength of a solution of sodium thiosulphate by means of a solution of sodium arsenite of known strength*

The reactions of iodine towards arsenious oxide and towards sodium thiosulphate being known, it is a simple matter to determine the strength of a solution of this latter substance by titration with a solution of iodine, the arsenic value of which is known. Accordingly the given solution of sodium thiosulphate was titrated with the same solution of iodine which had been used for titrating the arsenic solution in (2). Two titrations were made, 20 c.c. of the thiosulphate solution required 32·8 c.c. of iodine in each case.

In calculating the strength of the solution of sodium thiosulphate, it is to be borne in mind that 127 parts by weight of iodine react with 158 parts by weight of sodium thiosulphate (anhydrous) with formation of sodium iodide and tetrathionate, and that 4×127 parts by weight of iodine are capable of oxidizing 198 parts by weight of arsenious oxide. Consequently the equivalent weights of sodium thiosulphate and arsenious oxide are in the ratio of 4×158 to 198.

In the previous determination it was found that 20 c.c. of the solution of sodium arsenite containing 0·00481 grm. of As_2O_3 per c.c. required 32·1 c.c. of iodine solution. The volume of sodium thiosulphate solution equivalent to that volume of iodine is clearly $\frac{20 \times 32 \cdot 1}{32 \cdot 8}$ c.c. or 19·57 c.c. Denoting by y the weight of $Na_2S_2O_3$ per c.c., we determine y by solving the equation

$$\frac{19 \cdot 57 y}{20 \times 0 \cdot 00481} = \frac{4 \times 158}{198},$$

from which $y = 0 \cdot 0157$ grm. of $Na_2S_2O_3$ per c.c.

(4) *Determination of the strength of a solution of potassium dichromate iodometrically*

The solution of potassium dichromate, the strength of which was determined by means of recrystallized ferrous ammonium sulphate in (1), was treated with excess of potassium iodide in presence of a little dilute sulphuric acid and the liberated iodine titrated by means of standard sodium thiosulphate. The sodium thiosulphate solution was that standardized in (3). As a mean of several concordant titrations, it was found that 20 c.c. of the dichromate solution required 20·2 c.c. of thiosulphate. The strength of the solution of potassium dichromate x in grms. per c.c. was therefore obtained by solving the equation

$$\frac{20x}{20\cdot2 \times 0\cdot0157} = \frac{294}{6 \times 158},$$

from which $x = 0\cdot00492$ grm. of $K_2Cr_2O_7$ per c.c.

Comparing this result with that obtained in (1) in which the strength of the potassium dichromate solution was determined by means of recrystallized ferrous ammonium sulphate, the difference approximates to one per cent. Taking into account the fact that the solution of sodium thiosulphate was standardized by a somewhat indirect method, the agreement between the iron method and the iodometric method of determining the strength of the dichromate solution must be regarded as satisfactory.

(5) *An experiment to illustrate the relative magnitude of the errors involved by differences in the relative volumes of the measuring vessels*

In this experiment the strength of a solution of iodine was determined by means of the standard sodium thiosulphate in (3) placing first the iodine in the pipette and the thiosulphate in the burette, and secondly with the thiosulphate in the pipette and the iodine in the burette. The relative volumes of the pipette and of measured volumes from the burette were then determined by weighing the volumes of water delivered from these vessels. The following were the results obtained:

20 c.c. of iodine measured out by the pipette required as a mean result 20·3 c.c. of thiosulphate.

From this determination, the strength of the iodine equals 12·8 grms. per litre.

20 c.c. of sodium thiosulphate measured out by the pipette required 20·0 c.c. of iodine from the burette.

This determination gives the strength of the iodine solution as 12·6 grms. of iodine per litre. It is clear that the agreement between the two determinations is of the order of a difference of one-and-one-half per cent. This difference cannot possibly be due to any difficulty in carrying out the experiments, as the reaction between iodine and sodium thiosulphate is one of the most accurate reactions with which we have to deal, and the end-point is particularly easy to determine. It was therefore of interest to compare the relative volumes of the 20 c.c. pipette and of 20 c.c. of the burette.

It was found as a mean of several concordant measurements that the weight of water delivered from the pipette was 20·12 grms., while the weight of water delivered by running out 20 c.c. of water from the burette (taken between different points) was 19·92 grms. This difference was two parts in 200, or in other words the volume of the pipette was approximately one per cent. greater than that of an apparently equal volume of the burette.

(6) *Standardization of a solution of hydrochloric acid by means of sodium*

The hydrochloric acid was prepared so as to approximate to normal strength. A weighed piece of freshly cut sodium was dissolved in alcohol, and after the metal had been completely dissolved, excess of water was added to the solution. The solution thus obtained was next titrated with the hydrochloric acid, methyl red being used as indicator.

In the first experiment 0·454 grm. of sodium required 19·7 c.c. of hydrochloric acid.

Denoting by x the weight of hydrogen chloride in grms. per c.c. we obtain x by solution of the equation

$$\frac{19 \cdot 7x}{0 \cdot 454} = \frac{36 \cdot 5}{23},$$

from which $x = 0 \cdot 0366$ grm. HCl per c.c.

In the second experiment 0·659 grm. of sodium required 28·6 c.c. of hydrochloric acid.   From which we have

$$\frac{28\cdot6x}{0\cdot659} = \frac{36\cdot5}{23},$$

or $x = 0\cdot0366$ grm. HCl per c.c.

The agreement between the two experiments is perfect. It is clear that the hydrochloric acid is very nearly of normal strength, the normal solution containing 0·0365 grm. HCl per c.c.

(7) *Determination of the strength of a solution of potassium hydroxide by means of standard hydrochloric acid*

The hydrochloric acid the strength of which was determined by sodium in (6) was employed for titrating the solution of potash, methyl orange being employed as indicator.

In two experiments it was found that 20 c.c. of KOH required 19·5 c.c. of standard HCl.

Let y denote the strength of the solution of potash in grms. per c.c.   Then we have

$$\frac{20y}{19\cdot5 \times 0\cdot0366} = \frac{56}{36\cdot5},$$

from which $y = 0\cdot0547$ grm. per c.c.

(8) *Determination of the strength of a solution of potassium permanganate by means of a standard solution of potassium dichromate*

A solution of ferrous sulphate containing sulphuric acid was prepared, and aliquot portions of this solution were titrated in turn with the standard solution of potassium dichromate in (1) and with the solution of potassium permanganate.   As a result of several concordant experiments it was found that 25 c.c. of the iron solution required 24·6 c.c. of potassium dichromate and 24·3 c.c. of potassium permanganate.

It is clear that $24\cdot6 \times 0\cdot00496$ grm. of potassium dichromate is equivalent in oxidizing power to $24\cdot3x$ grms. of potassium permanganate, where x denotes the weight of potassium permanganate in grms. per c.c.   It is also an experimental fact that 316 parts by weight of potassium permanganate are equivalent to 490 parts by weight of potassium dichromate in oxidizing power. The value of x is therefore determined from the equation

$$\frac{24\cdot3x}{24\cdot6 \times 0\cdot00496} = \frac{316}{490},$$

from which $x = 0\cdot00324$ grm. $KMnO_4$ per c.c.

(9) Determination of the strength of a solution of oxalic acid by means of standard potash

Since oxalic acid is a weak acid and potassium hydroxide a strong base, phenolphthalein was employed as the indicator. The solution of potash in (7) was diluted to one-tenth of its original strength, making the strength $0\cdot00547$ grm. per c.c. The mean result of two concordant titrations was that 20 c.c. of the oxalic acid required the addition of 23·35 c.c. of potash,

$$\frac{20x}{23\cdot35 \times 0\cdot00547} = \frac{45}{56},$$

from which $x = 0\cdot00513$ grm. of $\begin{matrix} COOH \\ | \\ COOH \end{matrix}$ per c.c.

(10) Determination of the strength of the solution of the oxalic acid in (9) by means of standard potassium permanganate

The permanganate was that standardized in (8). Aliquot portions of the solution of oxalic acid were taken, a small quantity of sulphuric acid added to each, and each solution was heated before running in the permanganate. In the first experiment, the volume of permanganate required for complete oxidation of 20 c.c. of the solution of oxalic acid was 22·5 c.c., in the second experiment it was 22·4 c.c. From these results we have

$$\frac{20x}{22\cdot45 \times 0\cdot00324} = \frac{5 \times 90}{316},$$

from which $x = 0\cdot00517$ grm. of $\begin{matrix} COOH \\ | \\ COOH \end{matrix}$ per c.c.

The difference between the two results in (9) and (10) is less than one per cent., and this order of accuracy is as good as can be expected when it is borne in mind that both the potassium hydroxide and permanganate were standardized in a somewhat indirect manner.

(11) *Determination of the strength of a solution of silver nitrate by means of standard hydrochloric acid*

The hydrochloric acid prepared in (6) was diluted to one-tenth of its original strength as the solution of silver nitrate was prepared so as to be of approximately deci-normal strength. Aliquot portions of the solution of hydrochloric acid were measured out, and excess of calcium carbonate added to each. The contents of each flask were then titrated by means of silver nitrate, potassium chromate being employed as an indicator. As the mean of two concordant titrations it was found that 20 c.c. of the diluted hydrochloric acid required the addition of 20·25 c.c. of silver nitrate for complete precipitations. The weight x in grms. of silver nitrate in each c.c. of the given solution is therefore determined by the equation

$$\frac{20\cdot25x}{20 \times 0\cdot00366} = \frac{170}{36\cdot5},$$

from which $x = 0\cdot0168$ grm. $AgNO_3$ per c.c.

(12) *Determination of the amounts of hydrochloric acid and oxalic acid in a solution of the mixture*

This mixture was prepared by adding 500 c.c. of the solution of hydrochloric acid in (6) to 2000 c.c. of the solution of oxalic acid in (9) and (10). Determinations of the specific gravities of the constituent solutions and also of the resulting mixture showed that the total volume was very nearly 2500 c.c.; the difference between the observed and calculated values of the specific gravity of the mixture being negligible. The calculated values of the two acids in the mixture were therefore 7·32 grms. of hydrochloric acid and 4·12 grms. of oxalic acid (anhydrous) per litre.

This determination was carried out by titrating the hydrochloric acid by means of standard silver nitrate after addition of excess of calcium carbonate to the solution, potassium chromate being employed as an indicator. Then a further quantity of the mixture was measured out, and the mixed acids titrated by means of standard potassium hydroxide using phenolphthalein as indicator. The first experiment determined the weight of hydrochloric acid in a given volume of the mixture*. From the quantity of potash which is required to neutralize a definite volume of the mixed acids, it is possible to determine how much of the potash has been employed in the neutralization of the hydrochloric acid alone. The difference between the amount of potash which has been used up in the neutralization of the hydrochloric acid and that which has been required for the neutralization of the total acid is clearly equal to the amount which has been employed in the neutralization of the oxalic acid.

The standard solution of potash was diluted to one-fifth of its original strength making its new strength 0·0109 grm. per c.c.

It was found that 20 c.c. of the mixed acids required as a mean result 30·2 c.c. of potash and also that 20 c.c. of the mixture required 40·35 c.c. of standard silver nitrate.

From the result of the silver nitrate experiment it was found that the solution contained 7·29 grms. of hydrochloric acid per litre.

The weight of potassium hydroxide required to neutralize 7·29 grms. of hydrogen chloride is clearly equal to $\dfrac{56 \times 7\cdot29}{36\cdot5}$ grms. or 11·2 grms.

The weight of potash required to neutralize one litre of the mixture $= 50 \times 30\cdot2 \times 0\cdot0109$ grms. or 16·5 grms.

Hence the weight of potash required for the neutralization of the oxalic acid alone equals (16·5 − 11·2) grms. or 5·3 grms.

Since 56 parts by weight of potassium hydroxide neutralize 45 parts by weight of oxalic acid (anhydrous), it is clear that the weight of oxalic acid neutralized by 5·3 grms. of potash $= \dfrac{45 \times 5\cdot3}{56}$ grms. or 4·26 grms.

The given solution contains therefore 7·29 grms. of hydrochloric acid and 4·26 grms. of oxalic acid (anhydrous) per litre.

---

* It is necessary to neutralize with calcium carbonate in order to precipitate the very sparingly soluble calcium oxalate, and thereby prevent the precipitation of silver oxalate.

Calculation of the weights of the two acids from a single titration with potash

The total weight of the mixed acids in one litre of the solution equals 11·44 grms.

Let x denote the weight of oxalic acid and y the weight of hydrochloric acid per litre, then

$$x + y = 11\text{·}44 \quad \ldots\ldots\ldots\ldots\ldots\ldots (1),$$

$$\frac{56x}{45} + \frac{56y}{36\text{·}5} = 30\text{·}2 \times 0\text{·}0109 \times 50 \quad \ldots\ldots\ldots (2).$$

Solution of these two simultaneous equations gave the following results:

$x = 3\text{·}62$ grms. of oxalic acid (anhydrous) per litre and

$y = 7\text{·}82$ grms. of hydrochloric acid per litre.

This example is sufficient to illustrate the limitations of the accuracy of the general method of determining the amounts of two constituents of a mixture when the total weight is known from the difference between the equivalent weights, a small error in the experimental work giving rise to a considerable error in the final result especially when the two constituents are not present in roughly equal proportions.

Abridged List of the Elements and their Atomic Weights (O = 16)

Element		Accurate Atomic Weight	Approximate Atomic Weight
Antimony	Sb	120·2	120
Arsenic	As	74·96	75
Boron	B	11·0	11
Bromine	Br	79·92	80
Cadmium	Cd	112·4	112
Calcium	Ca	40·07	40
Carbon	C	12·00	12
Chlorine	Cl	35·46	35·5
Chromium	Cr	52·0	52
Copper	Cu	63·57	63·6
Hydrogen	H	1·008	1
Iodine	I	126·92	127
Iron	Fe	55·84	56
Manganese	Mn	54·93	55
Nitrogen	N	14·01	14
Oxygen	O	16·00	16
Phosphorus	P	31·04	31
Potassium	K	39·1	39
Silver	Ag	107·88	108
Sodium	Na	23·0	23
Sulphur	S	32·07	32
Tin	Sn	119·0	119
Titanium	Ti	48·1	48
Uranium	U	238·5	238·5
Zinc	Zn	65·37	65·4

Formula Values of Certain Substances (O = 16)

Acetic acid, CH_3COOH, 60·03
Ammonia, NH_3, 17·034
Ammonium chloride, NH_4Cl, 53·5
Ammonium thiocyanate, NH_4CNS, 76·12
Antimonious oxide, Sb_2O_3, 288·4
Arsenious oxide, As_2O_3, 197·92

Borax, $Na_2B_4O_710H_2O$, 382·16

Calcium carbonate, $CaCO_3$, 100·07
Citric acid, CH_2COOH
$\quad\quad\quad\; | $
$\quad\quad\; C(OH)COOH + H_2O$, 210·08
$\quad\quad\quad\; |$
$\quad\quad\; CH_2COOH$
Cupric sulphate, $CuSO_45H_2O$, 249·67

Ferrous ammonium sulphate, $FeSO_4$ $(NH_4)_2SO_46H_2O$, 392·17. This substance contains one-seventh of its own weight of iron
Formaldehyde, 30·02

Hydrochloric acid, HCl, 36·47
Hydrogen peroxide, H_2O_2, 34·016
Hydrogen sulphide, H_2S, 34·09

Nitric acid, HNO_3, 63·02

Orthophosphoric acid, H_3PO_4, 98·064
Oxalic acid, $COOH$
$\quad\quad\quad | \quad$, $2H_2O$, 126·05
$\quad\quad\; COOH$

Potassium bromide, KBr, 119·03
Potassium chloride, KCl, 74·56
Potassium cyanide, KCN, 65·11

Potassium dichromate, $K_2Cr_2O_7$, 294·2
Potassium ferrocyanide,$K_4Fe(CN)_63H_2O$, 422·36
Potassium hydroxide, KOH, 56·1
Potassium iodide, KI, 166·02
Potassium permanganate, $KMnO_4$, 158·03

Silver nitrate, $AgNO_3$, 169·89
Sodium bicarbonate, 84·01
Sodium carbonate (anhydrous), Na_2CO_3, 106
Sodium carbonate, decahydrate, $Na_2CO_310H_2O$, 286·16
Sodium chloride, $NaCl$, 58·46
Sodium hydroxide, $NaOH$, 40·01
Sodium thiosulphate, $Na_2S_2O_35H_2O$, 248·22
Succinic acid, CH_2COOH, 118·05
$\quad\quad\quad\quad | $
$\quad\quad\quad CH_2COOH$
Sulphur dioxide, SO_2, 64·07
Sulphuric acid, H_2SO_4, 98·086

Tartaric acid, $CH(OH)COOH$, 150·05
$\quad\quad\quad\quad\quad | $
$\quad\quad\quad\; CH(OH)COOH$

Water, H_2O, 18·016 $\quad$ $7H_2O$, 126·11
$\quad\quad\; 2H_2O$, 36·032 $\quad$ $8H_2O$, 144·13
$\quad\quad\; 3H_2O$, 54·048 $\quad$ $9H_2O$, 162·14
$\quad\quad\; 4H_2O$, 72·064 $\quad$ $10H_2O$, 180·16
$\quad\quad\; 5H_2O$, 90·08 $\quad\;$ $11H_2O$, 198·18
$\quad\quad\; 6H_2O$, 108·10 $\quad$ $12H_2O$, 216·19

Zinc sulphate, $ZnSO_47H_2O$, 287·55

Table of Hardness, Parts in 100,000

Column I denotes the number of c.c. of standard soap solution, and Column II the amount of $CaCO_3$ per 100,000

I	II	I	II	I	II	I	II	I	H
0·7	0·00	3·8	4·29	6·9	8·71	10·0	13·31	13·1	18·17
0·8	·16	·9	·43	7·0	·86	·1	·46	·2	·33
0·9	·32	4·0	·57	·1	9·00	·2	·61	·3	·49
1·0	·48	·1	·71	·2	·14	·3	·76	·4	·65
·1	·63	·2	·86	·3	·29	·4	·91	·5	·81
·2	·79	·3	5·00	·4	·43	·5	14·06	·6	·97
·3	·95	·4	·14	·5	·57	·6	·21	·7	19·13
·4	1·11	·5	·29	·6	·71	·7	·37	·8	·29
·5	·27	·6	·43	·7	·86	·8	·52	·9	·44
·6	·43	·7	·57	·8	10·00	·9	·68	14·0	·60
·7	·56	·8	·71	·9	·15	11·0	·84	·1	·76
·8	·69	·9	·86	8·0	·30	·1	15·00	·2	·92
·9	·82	5·0	6·00	·1	·45	·2	·16	·3	20·08
2·0	·95	·1	·14	·2	·60	·3	·32	·4	·24
·1	2·08	·2	·29	·3	·75	·4	·48	·5	·40
·2	·21	·3	·43	·4	·90	·5	·63	·6	·56
·3	·34	·4	·57	·5	11·05	·6	·79	·7	·71
·4	·47	·5	·71	·6	·20	·7	·95	·8	·87
·5	·60	·6	·86	·7	·35	·8	16·11	·9	21·03
·6	·73	·7	7·00	·8	·50	·9	·27	15·0	·19
·7	·86	·8	·14	·9	·65	12·0	·43	·1	·35
·8	·99	·9	·29	9·0	·80	·1	·59	·2	·51
·9	3·12	6·0	·43	·1	·95	·2	·75	·3	·68
3·0	·25	·1	·57	·2	12·11	·3	·90	·4	·85
·1	·38	·2	·71	·3	·26	·4	17·06	·5	22·02
·2	·51	·3	·86	·4	·41	·5	·22	·6	·18
·3	·64	·4	8·00	·5	·56	·6	·38	·7	·35
·4	·77	·5	·14	·6	·71	·7	·54	·8	·52
·5	·90	·6	·29	·7	·86	·8	·70	·9	·69
·6	4·03	·7	·43	·8	13·01	·9	·86	16·0	·86
·7	·16	·8	·57	·9	·16	13·0	18·02		

Relation between specific gravity and percentage composition of solutions of hydrochloric acid (in grammes of solute to 100 grammes of solution)

(Lunge and Marchlewski)

d_4^{15}	%	d_4^{15}	%	d_4^{15}	%
1·000	0·16	1·070	14·17	1·140	27·66
1·005	1·15	1·075	15·16	1·145	28·61
1·010	2·14	1·080	16·15	1·150	29·57
1·015	3·12	1·085	17·13	1·155	30·55
1·020	4·13	1·090	18·11	1·160	31·52
1·025	5·15	1·095	19·06	1·165	32·49
1·030	6·15	1·100	20·01	1·170	33·46
1·035	7·15	1·105	20·97	1·175	34·42
1·040	8·16	1·110	21·92	1·180	35·39
1·045	9·16	1·115	22·86	1·185	36·31
1·050	10·17	1·120	23·82	1·190	37·23
1·055	11·18	1·125	24·78	1·195	38·16
1·060	12·19	1·130	25·75	1·200	39·11
1·065	13·19	1·135	26·70		

Relation between specific gravity and percentage composition of solutions of sodium hydroxide (in grammes of solute to 100 grammes of solution)

Bousfield and Lowry (*Phil. Trans.*, 1905, 204 A, p. 253)

d_4^{15}	%	d_4^{15}	%	d_4^{15}	%
0·99918	0	1·18868	17	1·3728	34
1·01065	1	1·19973	18	1·3830	35
1·02198	2	1·21079	19	1·3933	36
1·03322	3	1·22183	20	1·4034	37
1·04441	4	1·23285	21	1·4135	38
1·05554	5	1·24386	22	1·4235	39
1·06666	6	1·25485	23	1·4334	40
1·07777	7	1·26582	24	1·4432	41
1·08887	8	1·27679	25	1·4529	42
1·09997	9	1·2877	26	1·4625	43
1·11107	10	1·2986	27	1·4720	44
1·12217	11	1·3094	28	1·4815	45
1·13327	12	1·3202	29	1·4911	46
1·14436	13	1·3309	30	1·5007	47
1·15545	14	1·3415	31	1·5102	48
1·16653	15	1·3520	32	1·5196	49
1·17761	16	1·3624	33	1·5290	50

LOGARITHMS.

	0	1	2	3	4	5	6	7	8	9	1	2	3	4	5	6	7	8	9
10	˙0000	0043	0086	0128	0170	0212	0253	0294	0334	0374	4	8	12	17	21	25	29	33	37
11	˙0414	0453	0492	0531	0569	0607	0645	0682	0719	0755	4	8	11	15	19	23	26	30	34
12	˙0792	0828	0864	0899	0934	0969	1004	1038	1072	1106	3	7	10	14	17	21	24	28	31
13	˙1139	1173	1206	1239	1271	1303	1335	1367	1399	1430	3	6	10	13	16	19	23	26	29
14	˙1461	1492	1523	1553	1584	1614	1644	1673	1703	1732	3	6	9	12	15	18	21	24	27
15	˙1761	1790	1818	1847	1875	1903	1931	1959	1987	2014	3	6	8	11	14	17	20	22	25
16	˙2041	2068	2095	2122	2148	2175	2201	2227	2253	2279	3	5	8	11	13	16	18	21	24
17	˙2304	2330	2355	2380	2405	2430	2455	2480	2504	2529	2	5	7	10	12	15	17	20	22
18	˙2553	2577	2601	2625	2648	2672	2695	2718	2742	2765	2	5	7	9	12	14	16	19	21
19	˙2788	2810	2833	2856	2878	2900	2923	2945	2967	2989	2	4	7	9	11	13	16	18	20
20	˙3010	3032	3054	3075	3096	3118	3139	3160	3181	3201	2	4	6	8	11	13	15	17	19
21	˙3222	3243	3263	3284	3304	3324	3345	3365	3385	3404	2	4	6	8	10	12	14	16	18
22	˙3424	3444	3464	3483	3502	3522	3541	3560	3579	3598	2	4	6	8	10	12	14	15	17
23	˙3617	3636	3655	3674	3692	3711	3729	3747	3766	3784	2	4	6	7	9	11	13	15	17
24	˙3802	3820	3838	3856	3874	3892	3909	3927	3945	3962	2	4	5	7	9	11	12	14	16
25	˙3979	3997	4014	4031	4048	4065	4082	4099	4116	4133	2	3	5	7	9	10	12	14	15
26	˙4150	4166	4183	4200	4216	4232	4249	4265	4281	4298	2	3	5	7	8	10	11	13	15
27	˙4314	4330	4346	4362	4378	4393	4409	4425	4440	4456	2	3	5	6	8	9	11	13	14
28	˙4472	4487	4502	4518	4533	4548	4564	4579	4594	4609	2	3	5	6	8	9	11	12	14
29	˙4624	4639	4654	4669	4683	4698	4713	4728	4742	4757	1	3	4	6	7	9	10	12	13
30	˙4771	4786	4800	4814	4829	4843	4857	4871	4886	4900	1	3	4	6	7	9	10	11	13
31	˙4914	4928	4942	4955	4969	4983	4997	5011	5024	5038	1	3	4	6	7	8	10	11	12
32	˙5051	5065	5079	5092	5105	5119	5132	5145	5159	5172	1	3	4	5	7	8	9	11	12
33	˙5185	5198	5211	5224	5237	5250	5263	5276	5289	5302	1	3	4	5	6	8	9	10	12
34	˙5315	5328	5340	5353	5366	5378	5391	5403	5416	5428	1	3	4	5	6	8	9	10	11
35	˙5441	5453	5465	5478	5490	5502	5514	5527	5539	5551	1	2	4	5	6	7	9	10	11
36	˙5563	5575	5587	5599	5611	5623	5635	5647	5658	5670	1	2	4	5	6	7	8	10	11
37	˙5682	5694	5705	5717	5729	5740	5752	5763	5775	5786	1	2	3	5	6	7	8	9	10
38	˙5798	5809	5821	5832	5843	5855	5866	5877	5888	5899	1	2	3	5	6	7	8	9	10
39	˙5911	5922	5933	5944	5955	5966	5977	5988	5999	6010	1	2	3	4	5	7	8	9	10
40	˙6021	6031	6042	6053	6064	6075	6085	6096	6107	6117	1	2	3	4	5	6	8	9	10
41	˙6128	6138	6149	6160	6170	6180	6191	6201	6212	6222	1	2	3	4	5	6	7	8	9
42	˙6232	6243	6253	6263	6274	6284	6294	6304	6314	6325	1	2	3	4	5	6	7	8	9
43	˙6335	6345	6355	6365	6375	6385	6395	6405	6415	6425	1	2	3	4	5	6	7	8	9
44	˙6435	6444	6454	6464	6474	6484	6493	6503	6513	6522	1	2	3	4	5	6	7	8	9
45	˙6532	6542	6551	6561	6571	6580	6590	6599	6609	6618	1	2	3	4	5	6	7	8	9
46	˙6628	6637	6646	6656	6665	6675	6684	6693	6702	6712	1	2	3	4	5	6	7	7	8
47	˙6721	6730	6739	6749	6758	6767	6776	6785	6794	6803	1	2	3	4	5	5	6	7	8
48	˙6812	6821	6830	6839	6848	6857	6866	6875	6884	6893	1	2	3	4	4	5	6	7	8
49	˙6902	6911	6920	6928	6937	6946	6955	6964	6972	6981	1	2	3	4	4	5	6	7	8
50	˙6990	6998	7007	7016	7024	7033	7042	7050	7059	7067	1	2	3	3	4	5	6	7	8
51	˙7076	7084	7093	7101	7110	7118	7126	7135	7143	7152	1	2	3	3	4	5	6	7	8
52	˙7160	7168	7177	7185	7193	7202	7210	7218	7226	7235	1	2	2	3	4	5	6	7	7
53	˙7243	7251	7259	7267	7275	7284	7292	7300	7308	7316	1	2	2	3	4	5	6	6	7
54	˙7324	7332	7340	7348	7356	7364	7372	7380	7388	7396	1	2	2	3	4	5	6	6	7

	0	1	2	3	4	5	6	7	8	9	1	2	3	4	5	6	7	8	9
55	·7404	7412	7419	7427	7435	7443	7451	7459	7466	7474	1	2	2	3	4	5	5	6	7
56	·7482	7490	7497	7505	7513	7520	7528	7536	7543	7551	1	2	2	3	4	5	5	6	7
57	·7559	7566	7574	7582	7589	7597	7604	7612	7619	7627	1	2	2	3	4	5	5	6	7
58	·7634	7642	7649	7657	7664	7672	7679	7686	7694	7701	1	1	2	3	4	4	5	6	7
59	·7709	7716	7723	7731	7738	7745	7752	7760	7767	7774	1	1	2	3	4	4	5	6	7
60	·7782	7789	7796	7803	7810	7818	7825	7832	7839	7846	1	1	2	3	4	4	5	6	6
61	·7853	7860	7868	7875	7882	7889	7896	7903	7910	7917	1	1	2	3	4	4	5	6	6
62	·7924	7931	7938	7945	7952	7959	7966	7973	7980	7987	1	1	2	3	3	4	5	6	6
63	·7993	8000	8007	8014	8021	8028	8035	8041	8048	8055	1	1	2	3	3	4	5	5	6
64	·8062	8069	8075	8082	8089	8096	8102	8109	8116	8122	1	1	2	3	3	4	5	5	6
65	·8129	8136	8142	8149	8156	8162	8169	8176	8182	8189	1	1	2	3	3	4	5	5	6
66	·8195	8202	8209	8215	8222	8228	8235	8241	8248	8254	1	1	2	3	3	4	5	5	6
67	·8261	8267	8274	8280	8287	8293	8299	8306	8312	8319	1	1	2	3	3	4	5	5	6
68	·8325	8331	8338	8344	8351	8357	8363	8370	8376	8382	1	1	2	3	3	4	4	5	6
69	·8388	8395	8401	8407	8414	8420	8426	8432	8439	8445	1	1	2	2	3	4	4	5	6
70	·8451	8457	8463	8470	8476	8482	8488	8494	8500	8506	1	1	2	2	3	4	4	5	6
71	·8513	8519	8525	8531	8537	8543	8549	8555	8561	8567	1	1	2	2	3	4	4	5	5
72	·8573	8579	8585	8591	8597	8603	8609	8615	8621	8627	1	1	2	2	3	4	4	5	5
73	·8633	8639	8645	8651	8657	8663	8669	8675	8681	8686	1	1	2	2	3	4	4	5	5
74	·8692	8698	8704	8710	8716	8722	8727	8733	8739	8745	1	1	2	2	3	4	4	5	5
75	·8751	8756	8762	8768	8774	8779	8785	8791	8797	8802	1	1	2	2	3	3	4	5	5
76	·8808	8814	8820	8825	8831	8837	8842	8848	8854	8859	1	1	2	2	3	3	4	5	5
77	·8865	8871	8876	8882	8887	8893	8899	8904	8910	8915	1	1	2	2	3	3	4	4	5
78	·8921	8927	8932	8938	8943	8949	8954	8960	8965	8971	1	1	2	2	3	3	4	4	5
79	·8976	8982	8987	8993	8998	9004	9009	9015	9020	9025	1	1	2	2	3	3	4	4	5
80	·9031	9036	9042	9047	9053	9058	9063	9069	9074	9079	1	1	2	2	3	3	4	4	5
81	·9085	9090	9096	9101	9106	9112	9117	9122	9128	9133	1	1	2	2	3	3	4	4	5
82	·9138	9143	9149	9154	9159	9165	9170	9175	9180	9186	1	1	2	2	3	3	4	4	5
83	·9191	9196	9201	9206	9212	9217	9222	9227	9232	9238	1	1	2	2	3	3	4	4	5
84	·9243	9248	9253	9258	9263	9269	9274	9279	9284	9289	1	1	2	2	3	3	4	4	5
85	·9294	9299	9304	9309	9315	9320	9325	9330	9335	9340	1	1	2	2	3	3	4	4	5
86	·9345	9350	9355	9360	9365	9370	9375	9380	9385	9390	1	1	1	2	3	3	4	4	5
87	·9395	9400	9405	9410	9415	9420	9425	9430	9435	9440	0	1	1	2	2	3	3	4	4
88	·9445	9450	9455	9460	9465	9469	9474	9479	9484	9489	0	1	1	2	2	3	3	4	4
89	·9494	9499	9504	9509	9513	9518	9523	9528	9533	9538	0	1	1	2	2	3	3	4	4
90	·9542	9547	9552	9557	9562	9566	9571	9576	9581	9586	0	1	1	2	2	3	3	4	4
91	·9590	9595	9600	9605	9609	9614	9619	9624	9628	9633	0	1	1	2	2	3	3	4	4
92	·9638	9643	9647	9652	9657	9661	9666	9671	9675	9680	0	1	1	2	2	3	3	4	4
93	·9685	9689	9694	9699	9703	9708	9713	9717	9722	9727	0	1	1	2	2	3	3	4	4
94	·9731	9736	9741	9745	9750	9754	9759	9763	9768	9773	0	1	1	2	2	3	3	4	4
95	·9777	9782	9786	9791	9795	9800	9805	9809	9814	9818	0	1	1	2	2	3	3	4	4
96	·9823	9827	9832	9836	9841	9845	9850	9854	9859	9863	0	1	1	2	2	3	3	4	4
97	·9868	9872	9877	9881	9886	9890	9894	9899	9903	9908	0	1	1	2	2	3	3	4	4
98	·9912	9917	9921	9926	9930	9934	9939	9943	9948	9952	0	1	1	2	2	3	3	4	4
99	·9956	9961	9965	9969	9974	9978	9983	9987	9991	9996	0	1	1	2	2	3	3	3	4

ANTI-LOGARITHMS.

	0	1	2	3	4	5	6	7	8	9	1	2	3	4	5	6	7	8	9
·00	1000	1002	1005	1007	1009	1012	1014	1016	1019	1021	0	0	1	1	1	1	2	2	2
·01	1023	1026	1028	1030	1033	1035	1038	1040	1042	1045	0	0	1	1	1	1	2	2	2
·02	1047	1050	1052	1054	1057	1059	1062	1064	1067	1069	0	0	1	1	1	1	2	2	2
·03	1072	1074	1076	1079	1081	1084	1086	1089	1091	1094	0	0	1	1	1	1	2	2	2
·04	1096	1099	1102	1104	1107	1109	1112	1114	1117	1119	0	1	1	1	1	2	2	2	2
·05	1122	1125	1127	1130	1132	1135	1138	1140	1143	1146	0	1	1	1	1	2	2	2	2
·06	1148	1151	1153	1156	1159	1161	1164	1167	1169	1172	0	1	1	1	1	2	2	2	2
·07	1175	1178	1180	1183	1186	1189	1191	1194	1197	1199	0	1	1	1	1	2	2	2	2
·08	1202	1205	1208	1211	1213	1216	1219	1222	1225	1227	0	1	1	1	1	2	2	2	3
·09	1230	1233	1236	1239	1242	1245	1247	1250	1253	1256	0	1	1	1	1	2	2	2	3
·10	1259	1262	1265	1268	1271	1274	1276	1279	1282	1285	0	1	1	1	1	2	2	2	3
·11	1288	1291	1294	1297	1300	1303	1306	1309	1312	1315	0	1	1	1	2	2	2	2	3
·12	1318	1321	1324	1327	1330	1334	1337	1340	1343	1346	0	1	1	1	2	2	2	2	3
·13	1349	1352	1355	1358	1361	1365	1368	1371	1374	1377	0	1	1	1	2	2	2	3	3
·14	1380	1384	1387	1390	1393	1396	1400	1403	1406	1409	0	1	1	1	2	2	2	3	3
·15	1413	1416	1419	1422	1426	1429	1432	1435	1439	1442	0	1	1	1	2	2	2	3	3
·16	1445	1449	1452	1455	1459	1462	1466	1469	1472	1476	0	1	1	1	2	2	2	3	3
·17	1479	1483	1486	1489	1493	1496	1500	1503	1507	1510	0	1	1	1	2	2	3	3	3
·18	1514	1517	1521	1524	1528	1531	1535	1538	1542	1545	0	1	1	1	2	2	3	3	3
·19	1549	1552	1556	1560	1563	1567	1570	1574	1578	1581	0	1	1	1	2	2	3	3	3
·20	1585	1589	1592	1596	1600	1603	1607	1611	1614	1618	0	1	1	1	2	2	3	3	3
·21	1622	1626	1629	1633	1637	1641	1644	1648	1652	1656	0	1	1	2	2	2	3	3	3
·22	1660	1663	1667	1671	1675	1679	1683	1687	1690	1694	0	1	1	2	2	2	3	3	3
·23	1698	1702	1706	1710	1714	1718	1722	1726	1730	1734	0	1	1	2	2	2	3	3	4
·24	1738	1742	1746	1750	1754	1758	1762	1766	1770	1774	0	1	1	2	2	2	3	3	4
·25	1778	1782	1786	1791	1795	1799	1803	1807	1811	1816	0	1	1	2	2	2	3	3	4
·26	1820	1824	1828	1832	1837	1841	1845	1849	1854	1858	0	1	1	2	2	3	3	3	4
·27	1862	1866	1871	1875	1879	1884	1888	1892	1897	1901	0	1	1	2	2	3	3	3	4
·28	1905	1910	1914	1919	1923	1928	1932	1936	1941	1945	0	1	1	2	2	3	3	4	4
·29	1950	1954	1959	1963	1968	1972	1977	1982	1986	1991	0	1	1	2	2	3	3	4	4
·30	1995	2000	2004	2009	2014	2018	2023	2028	2032	2037	0	1	1	2	2	3	3	4	4
·31	2042	2046	2051	2056	2061	2065	2070	2075	2080	2084	0	1	1	2	2	3	3	4	4
·32	2089	2094	2099	2104	2109	2113	2118	2123	2128	2133	0	1	1	2	2	3	3	4	4
·33	2138	2143	2148	2153	2158	2163	2168	2173	2178	2183	0	1	1	2	2	3	3	4	4
·34	2188	2193	2198	2203	2208	2213	2218	2223	2228	2234	1	1	2	2	3	3	4	4	5
·35	2239	2244	2249	2254	2259	2265	2270	2275	2280	2286	1	1	2	2	3	3	4	4	5
·36	2291	2296	2301	2307	2312	2317	2323	2328	2333	2339	1	1	2	2	3	3	4	4	5
·37	2344	2350	2355	2360	2366	2371	2377	2382	2388	2393	1	1	2	2	3	3	4	4	5
·38	2399	2404	2410	2415	2421	2427	2432	2438	2443	2449	1	1	2	2	3	3	4	4	5
·39	2455	2460	2466	2472	2477	2483	2489	2495	2500	2506	1	1	2	2	3	3	4	5	5
·40	2512	2518	2523	2529	2535	2541	2547	2553	2559	2564	1	1	2	2	3	4	4	5	5
·41	2570	2576	2582	2588	2594	2600	2606	2612	2618	2624	1	1	2	2	3	4	4	5	5
·42	2630	2636	2642	2649	2655	2661	2667	2673	2679	2685	1	1	2	2	3	4	4	5	6
·43	2692	2698	2704	2710	2716	2723	2729	2735	2742	2748	1	1	2	3	3	4	4	5	6
·44	2754	2761	2767	2773	2780	2786	2793	2799	2805	2812	1	1	2	3	3	4	4	5	6
·45	2818	2825	2831	2838	2844	2851	2858	2864	2871	2877	1	1	2	3	3	4	5	5	6
·46	2884	2891	2897	2904	2911	2917	2924	2931	2938	2944	1	1	2	3	3	4	5	5	6
·47	2951	2958	2965	2972	2979	2985	2992	2999	3006	3013	1	1	2	3	3	4	5	5	6
·48	3020	3027	3034	3041	3048	3055	3062	3069	3076	3083	1	1	2	3	4	4	5	6	6
·49	3090	3097	3105	3112	3119	3126	3133	3141	3148	3155	1	1	2	3	4	4	5	6	6

	0	1	2	3	4	5	6	7	8	9	1	2	3	4	5	6	7	8	9
50	3162	3170	3177	3184	3192	3199	3206	3214	3221	3228	1	1	2	3	4	4	5	6	7
·51	3236	3243	3251	3258	3266	3273	3281	3289	3296	3304	1	2	2	3	4	5	5	6	7
·52	3311	3319	3327	3334	3342	3350	3357	3365	3373	3381	1	2	2	3	4	5	5	6	7
·53	3388	3396	3404	3412	3420	3428	3436	3443	3451	3459	1	2	2	3	4	5	6	6	7
·54	3467	3475	3483	3491	3499	3508	3516	3524	3532	3540	1	2	2	3	4	5	6	6	7
·55	3548	3556	3565	3573	3581	3589	3597	3606	3614	3622	1	2	2	3	4	5	6	7	7
·56	3631	3639	3648	3656	3664	3673	3681	3690	3698	3707	1	2	3	3	4	5	6	7	8
·57	3715	3724	3733	3741	3750	3758	3767	3776	3784	3793	1	2	3	3	4	5	6	7	8
·58	3802	3811	3819	3828	3837	3846	3855	3864	3873	3882	1	2	3	4	4	5	6	7	8
·59	3890	3899	3908	3917	3926	3936	3945	3954	3963	3972	1	2	3	4	5	5	6	7	8
60	3981	3990	3999	4009	4018	4027	4036	4046	4055	4064	1	2	3	4	5	6	6	7	8
·61	4074	4083	4093	4102	4111	4121	4130	4140	4150	4159	1	2	3	4	5	6	7	8	9
·62	4169	4178	4188	4198	4207	4217	4227	4236	4246	4256	1	2	3	4	5	6	7	8	9
·63	4266	4276	4285	4295	4305	4315	4325	4335	4345	4355	1	2	3	4	5	6	7	8	9
·64	4365	4375	4385	4395	4406	4416	4426	4436	4446	4457	1	2	3	4	5	6	7	8	9
·65	4467	4477	4487	4498	4508	4519	4529	4539	4550	4560	1	2	3	4	5	6	7	8	9
·66	4571	4581	4592	4603	4613	4624	4634	4645	4656	4667	1	2	3	4	5	6	7	9	10
·67	4677	4688	4699	4710	4721	4732	4742	4753	4764	4775	1	2	3	4	5	7	8	9	10
·68	4786	4797	4808	4819	4831	4842	4853	4864	4875	4887	1	2	3	4	6	7	8	9	10
·69	4898	4909	4920	4932	4943	4955	4966	4977	4989	5000	1	2	3	5	6	7	8	9	10
70	5012	5023	5035	5047	5058	5070	5082	5093	5105	5117	1	2	4	5	6	7	8	9	11
·71	5129	5140	5152	5164	5176	5188	5200	5212	5224	5236	1	2	4	5	6	7	8	10	11
·72	5248	5260	5272	5284	5297	5309	5321	5333	5346	5358	1	2	4	5	6	7	9	10	11
·73	5370	5383	5395	5408	5420	5433	5445	5458	5470	5483	1	3	4	5	6	8	9	10	11
·74	5495	5508	5521	5534	5546	5559	5572	5585	5598	5610	1	3	4	5	6	8	9	10	12
·75	5623	5636	5649	5662	5675	5689	5702	5715	5728	5741	1	3	4	5	7	8	9	10	12
·76	5754	5768	5781	5794	5808	5821	5834	5848	5861	5875	1	3	4	5	7	8	9	11	12
·77	5888	5902	5916	5929	5943	5957	5970	5984	5998	6012	1	3	4	5	7	8	10	11	12
·78	6026	6039	6053	6067	6081	6095	6109	6124	6138	6152	1	3	4	6	7	8	10	11	13
·79	6166	6180	6194	6209	6223	6237	6252	6266	6281	6295	1	3	4	6	7	9	10	11	13
80	6310	6324	6339	6353	6368	6383	6397	6412	6427	6442	1	3	4	6	7	9	10	12	13
·81	6457	6471	6486	6501	6516	6531	6546	6561	6577	6592	2	3	5	6	8	9	11	12	14
·82	6607	6622	6637	6653	6668	6683	6699	6714	6730	6745	2	3	5	6	8	9	11	12	14
·83	6761	6776	6792	6808	6823	6839	6855	6871	6887	6902	2	3	5	6	8	9	11	13	14
·84	6918	6934	6950	6966	6982	6998	7015	7031	7047	7063	2	3	5	6	8	10	11	13	15
·85	7079	7096	7112	7129	7145	7161	7178	7194	7211	7228	2	3	5	7	8	10	12	13	15
·86	7244	7261	7278	7295	7311	7328	7345	7362	7379	7396	2	3	5	7	8	10	12	13	15
·87	7413	7430	7447	7464	7482	7499	7516	7534	7551	7568	2	3	5	7	9	10	12	14	16
·88	7586	7603	7621	7638	7656	7674	7691	7709	7727	7745	2	4	5	7	9	11	12	14	16
·89	7762	7780	7798	7816	7834	7852	7870	7889	7907	7925	2	4	5	7	9	11	13	14	16
90	7943	7962	7980	7998	8017	8035	8054	8072	8091	8110	2	4	6	7	9	11	13	15	17
·91	8128	8147	8166	8185	8204	8222	8241	8260	8279	8299	2	4	6	8	9	11	13	15	17
·92	8318	8337	8356	8375	8395	8414	8433	8453	8472	8492	2	4	6	8	10	12	14	15	17
·93	8511	8531	8551	8570	8590	8610	8630	8650	8670	8690	2	4	6	8	10	12	14	16	18
·94	8710	8730	8750	8770	8790	8810	8831	8851	8872	8892	2	4	6	8	10	12	14	16	18
·95	8913	8933	8954	8974	8995	9016	9036	9057	9078	9099	2	4	6	8	10	12	15	17	19
·96	9120	9141	9162	9183	9204	9226	9247	9268	9290	9311	2	4	6	8	11	13	15	17	19
·97	9333	9354	9376	9397	9419	9441	9462	9484	9506	9528	2	4	7	9	11	13	15	17	20
·98	9550	9572	9594	9616	9638	9661	9683	9705	9727	9750	2	4	7	9	11	13	16	18	20
·99	9772	9795	9817	9840	9863	9886	9908	9931	9954	9977	2	5	7	9	11	14	16	18	20

INDEX

Printed in the United States
By Bookmasters